PROJECT-BASED LEARNING
in the Math Classroom

Project-Based Learning in the Math Classroom: Grades K–2 explains how to keep inquiry at the heart of mathematics teaching in the elementary grades. Helping teachers integrate other subjects into the math classroom, this book outlines in-depth tasks, projects and routines to support Project-Based Learning (PBL). Featuring helpful tips for creating PBL units, alongside models and strategies that can be implemented immediately, *Project-Based Learning in the Math Classroom: Grades K–2* understands that teaching in a project-based environment means using great teaching practices. The authors impart strategies that assist teachers in planning standards-based lessons, encouraging wonder and curiosity, providing a safe environment where mistakes can occur and giving students opportunities for revision and reflection.

Telannia Norfar is a mathematics teacher at a public high school in Oklahoma City, OK. She has taught all high school courses including AP Calculus AB for over 15 years. As a former journalist and account manager, Telannia found Project-Based Learning a viable method for teaching worthy mathematical concepts.

Chris Fancher is a retired public school math and engineering teacher living in Round Rock, TX. He has taught every math course, from pre-algebra to calculus, in more than 20 years of public education as well as being a Project Lead the Way (PLTW) engineering teacher and an instructional coach for Project-Based Learning.

Books in the *Project-Based Learning in the Math Classroom* Collection

Available from Routledge
(www.routledge.com)

Project-Based Learning in the Math Classroom: Grades K–2
Telannia Norfar & Chris Fancher

Project-Based Learning in the Math Classroom: Grades 3–5
Telannia Norfar & Chris Fancher

Project-Based Learning in the Math Classroom: Grades 6–10
Telannia Norfar & Chris Fancher

PROJECT-BASED LEARNING

in the Math Classroom

Grades K–2

Telannia Norfar and

Chris Fancher

Routledge

Taylor & Francis Group

NEW YORK AND LONDON

Cover image: Shutterstock

First published 2022
by Routledge
605 Third Avenue, New York, NY 10158

and by Routledge
2 Park Square, Milton Park, Abingdon, Oxon, OX14 4RN

Routledge is an imprint of the Taylor & Francis Group, an informa business

© 2022 Taylor & Francis

Library of Congress Cataloging-in-Publication Data
Names: Norfar, Telannia, author. | Fancher, Chris, author.
Title: Project-based learning in the math classroom : grades K-2 / Telannia Norfar, Chris Fancher.
Description: New York : Routledge, 2022. | Includes bibliographical references. |
Identifiers: LCCN 2021044230 (print) | LCCN 2021044231 (ebook) | ISBN 9781646322114 (paperback) | ISBN 9781032145044 (hardback) | ISBN 9781003237358 (ebook)
Subjects: LCSH: Mathematics--Study and teaching (Elementary) | Project method in teaching.
Classification: LCC QA20.P73 N677 2022 (print) | LCC QA20.P73 (ebook) | DDC 372.7/044--dc23/eng/20211117
LC record available at https://lccn.loc.gov/2021044230
LC ebook record available at https://lccn.loc.gov/2021044231

ISBN: 978-1-032-14504-4 (hbk)
ISBN: 978-1-64632-211-4 (pbk)
ISBN: 978-1-003-23735-8 (ebk)

DOI: 10.4324/9781003237358

Typeset in Warnock Pro
by Deanta Global Publishing Services, Chennai, India

Access the Support Material: www.routledge.com/9781646322114.

Contents

Tables

Figures

Introduction

Telannia Norfar
Chris Fancher

In 2019, we published *Project-Based Learning in the Math Classroom Grades 6-10*. It was in response to constant requests for a book that showed people how to create projects with a math context. Quickly after we published the book, educators wanted an elementary version of the book, specifically one for K–2 and one for 3–5. When we decided we would write elementary versions of our book, we talked about our own educational experience and interactions with elementary teachers. I, Telannia, remembered my love of reading and doing logic puzzles. I spent hours reading or doing puzzles. My nightstand would have five to eight books as well as *Highlights* magazines. I loved reading and doing the puzzles in the magazine while curled up in my bed. When it comes to school, I have vivid memories of sitting on the mat in kindergarten as my teacher passionately read a book aloud. Report cards were filled with S for satisfactory which is the highest score given in the early 80s. I, Chris, have memories of treating my math experience as a contest in elementary school. We had leveled math books that were a different color for each grade and I always wanted to be the first one to complete each book. And, in class, we had timed math sheets. One hundred math problems in a certain amount of time. I always wanted to be the first to complete each one. Could you tell I was a bit competitive? But I couldn't tell you anything about what we did during math time. At home I always had books going. Like Telannia, I loved *Highlights* magazine and logic puzzles too. I loved to read and write in my English classes as well. Report cards were always high marks academically, but I was a bit rambunctious and my behavior grade wasn't the highest.

As we talked more and more, we realized from kindergarten to second grade our math experience was more outside of the classroom. Beyond counting numbers or memorizing math facts, we didn't have any math stories. However, we could remember counting money with our family or playing cards that involved so much mathematical understanding and joy. For decades when we interacted with elementary teachers, many K–2 teachers shared how they loved literacy more than math. Some shared they teach the lower-level grades because they don't think they could do the upper-level elementary math. This helped us to understand our own lack of vivid memories in elementary. Teachers also shared concerns of not having the time to do Project-Based Learning. These memories and discussions with educators shaped our focus for this version of our book.

For this book, we wanted to develop a love for mathematics and the ability to feel you can do Project-Based Learning (PBL). We want you to develop a love for mathematics because it is hard to do a project when you don't feel comfortable with the subject. Throughout the book, we hope to grow your comfort with the subject of mathematics. We provide an experience to help ground you in what it feels like to do a project based in math content. There are also practical explanations all throughout the book. Although the entire book is designed to develop a love, Chapter 4 specifically discusses the connections between language arts and mathematics. This helped us the most when we think about our own love of both subjects.

In addition to developing a love of math, projects are the greatest goal of this book. We know people are more likely to do Project-Based Learning if they feel like it is doable. The more it is like your current practice, the more you feel like you can do it. We worked to incorporate practices you already do in K–2. For example, K–2 teachers are great at using the "stages of play" (New Teacher Center). Play is a necessary component of mathematical learning. Jo Boaler, Jen Munson and Cathy Williams, authors of *Mindset Mathematics: Visualizing and Investigating Big Ideas, Grade K*, explains it is important to open mathematics by allowing students to play with mathematics (Boaler et al. p. 12–13). Teaching in a project-based environment means using great teaching practices like play. Many teachers can feel overwhelmed by all the researched based practices. However, we like to keep it simple. We see "great teaching practices" as strategies that involve the following:

(1) Planning lessons that are standards-based
(2) Encouraging wonder and curiosity
(3) Providing a safe environment where it is ok to make mistakes
(4) Giving students opportunity for revision and reflection

When considering the main components of a PBL-based classroom we see that "great" teachers are doing most of these practices already. What should be added would be a problem or issue that students need to take an interest in that leads to student inquiry and questioning; an opportunity for students to direct their learning and an opportunity to demonstrate their learning to an audience.

In this book, we provide in-depth tasks, projects and routines to support Project-Based Learning. It includes details about how to create a PBL unit as well as models that you can implement immediately. The book aligns with leading research available regarding inquiry-based learning methods. If you have chosen to read this book, you desire to help all students be great learners especially in the study of mathematics.

Thank you for joining us on this journey.

References

Boaler, Jo, et al. *Mindset Mathematics Grade K*. B Jossey-Bass, 2020.

New Teacher Center. "How to Enable Play in Remote Learning." *How to Enable Play in Remote Learning*, n.d., https://newteachercenter.org/blog/2021/02/25/enabling-play-in-remote-learning/.

How to Use This Book

Telannia Norfar
Chris Fancher

This book is the second in a three-part series on implementing Project-Based Learning in the math classroom. The first book, which is for grades 6–10, was designed in response to the following requests of teachers that the book would: be easy to read, explain how to create and implement a project with a math lens, and provide lots of examples. This book and the third (grades 3–5) complete the series and follow the same request with a couple of exceptions. Our first book operated on the concept that a teacher would only be teaching mathematics. In this book, we show the planning for integrating subjects but with math being the focus. We also understand mathematics is not always understood as well as literacy for K–2 teachers. As a result, we created a chapter on understanding mathematics through the lens of literacy. As a result, there was one chapter added and two combined from the original book.

Except for a change to Section II, we kept a similar format (style and chapter focus) to help with teachers' learning with others or as a school. Keeping a similar format allows teachers of different grade levels to purchase the book for their grade but discuss it easily with teachers from different grades because the formats are the same. It provides ease for the curriculum department or school coaches who may be supporting teachers using the books. For example, Chapter 2 is an experience. The situation is the same for the K–2 and the 3–5 book just in case you want to do a school-wide experience. However, some of the details in the experience are different to support the appropriate development that occurs in K–2 versus 3–5.

As teachers, we completely understand teachers' desire to make it easy to read. We know time is limited and the most helpful books are the ones that get to the point. We found skipping around can be a desirable trait for books. You can read the chapters in order or you can skip around to areas you need. The chapters work together within a section but can be considered individually. We worked really hard to provide visuals and concise information. If you are wanting to skip around, here is a synopsis of each chapter:

❏ Chapter 1: not sure if you should take the leap to do Projects-Based Learning especially for math? This chapter explains why PBL and math are the perfect match to help students understand mathematics. It provides lots of research-based evidence to support the need of projects being a part of a student's mathematical development.

❏ Chapter 2: you and colleagues or friends can experience a project to help you get a better understanding of projects by taking on the role of a student.

❏ Chapter 3: understand how to design a task or a project. There is also help on to what mathematics content lends itself to be the main content of a project or how it can be better to be taught with other strategies.

❏ Chapter 4: connections between literacy and math skills are explained to show how math is basically a literacy skill just with numbers instead of letters.

❏ Chapter 5: learn various strategies that are best utilized in and out of projects. Many of the strategies are focused on thinking and align with literacy skills.

❏ Chapter 6: this is an example of a project with complete details.

❏ Chapter 7: these are tasks for each grade level. Tasks are a reduced form of a project. It is typically completed in a couple of days and is similar to a performance task.

❏ Chapter 8: there are projects for each grade level but with just basic components like the project idea, standards and an example of a lesson.

UNDERSTANDING PROJECT-BASED LEARNING IN A MATH CLASSROOM

Mathematics and PBL – The Perfect Match

Between the two of us, we have been teaching mathematics for over 40 years. The entire time we have used Project-Based Learning as a tool to teach it. We have seen the difference it has made in our own students' lives. We know it is a structure that works regardless of grade level but we want to make the case why you should take this journey with us. In this chapter, we explain how the teaching and learning of mathematics for kindergarten to second grade is prime for Project-Based Learning. It is worth the time and helps you develop a comfort with mathematics if you are not comfortable now. Let's begin by looking at how it aligns with expectations of mathematical learning at the early-elementary level.

In 2020, the National Council of Teachers of Mathematics (NCTM) released *Catalyzing Change in Early Childhood and Elementary Mathematics* which stated, "each and every child can have access to learning environments that are designed as mathematically powerful spaces" (National Council of Teachers of Mathematics p. 1). These are spaces where students learn mathematics deeply and richly. The book was released out of a need for elementary educators to incorporate practices that support all students and to help students make sense of mathematics. The book outlines several needs but we want to highlight three of them as they best show how Project-Based Learning is the tool to use when learning mathematics. The three focus areas are: purpose of mathematics, equitable structures and equitable instruction. To ease reading

DOI: 10.4324/9781003237358-2

and allow for skipping around, the focus areas are divided into headings. Let's start with the purpose of mathematics.

Purpose of Mathematics

For years, the NCTM has been helping to shape mathematics education. Its work helped inform the Common Core State Standards (CCSS) which are still used by many states. States that don't use the Common Core still reference the NCTM's work to inform their state standards. When using the NCTM's framework, the purpose of mathematics is clear and aligns with the development of five- to seven-year-olds and their deep understanding, critique of the world and enjoyment of mathematics. This is exactly how we feel about our relationship with mathematics. And we are not alone. Other practitioners have written great articles or books that share the same, especially when it comes to deep understanding. Here are a couple highlighted for your ease of reading:

> Mike Flynn states in *Beyond Answers: Exploring Mathematical Practices with Young Children*, "The essential premise behind all of these efforts is to encourage and support math instruction that develops conceptual understanding with procedural proficiency."
>
> (Flynn p. 8)

> Stanford University Math Professor Jo Boaler in her Youcubed .org article called *Fluency Without Fear* states, "Research tells us that the best mathematics classrooms are those in which students learn number facts and number sense through engaging activities that focus on mathematical understanding rather than rote memorization."
>
> (Youcubed.org)

Deep understanding of mathematics is not just being able to skip count by 2s or 3s. It is understanding how we use numbers in different ways. For example, we use it for an amount of objects as well as distance. In Project-Based Learning, students are immersed for an extended period of time in a situation that develops conceptual and procedural understanding. For K–2, students' time to make sense of their world is what they need. They thrive when they come back to an idea over and over again. This leads us to the other purpose of mathematics which is to critique the world and find joy in it.

A beautiful aspect of the K–2 student is their natural ability to work to understand the world. They are filled with questions and examine the world around them naturally. This is exactly the heart of learning mathematics. The answer is not what is important; it is the question. In *Catalyzing Change*, the NCTM encourages teachers at the early childhood level to provide opportunities where students examine their own lives to make sense of mathematics (National Council of Teachers of Mathematics p. 16). It is through this examination that joy is found. It is where we found our own joy at the same age. Our homes were constantly helping us make sense of the quantitative world around us. We got excited when we figured out how to buy a new toy or how long it would take to get to our favorite place. Project-Based Learning requires students to make sense of the world using real-life situations or the desires of the students. However, we feel that this purpose cannot be fulfilled unless you have equity.

Equitable Structures

Equitable structures are when procedures and policies are in place that ensure all students receive rich mathematics instruction. For far too long rich learning experiences have not been equitable for students. We believe there are many reasons for this but two major ones are a lack of culturally responsive teaching as well as systematic issues such as tracking. In Lisa Delpit's *Multiplication is for White People: Raising Expectations for Other People's Children*, she lays out personal and researched work about the educational experience from kindergarten to university level for black children. She shares at the very beginning some reasons why black students are not achieving, despite tons of research showing their brilliance from birth. She states,

> The final reason I'd like to propose for why African American students are not achieving at levels commensurate with their ability has to do with curriculum content. If the curriculum we use to teach our children does not connect in positive ways to the culture young people bring to school, it is doomed to failure.
>
> (Delpit p. 21)

Project-Based Learning requires teachers to take into account their students. You must think about their background and interest and use it to develop projects. This is the heart of being culturally responsive which is an equitable structure. Even with teachers changing to be more culturally responsive, inequity can still happen by continuing some of our school policies.

In *Catalyzing Change*, the NCTM identified barriers that exclude or limit children's access and opportunity. One of the key policies that prevent equitable structures in mathematics is tracking. "Ability grouping and tracking of children lead to differential learning opportunities that not only widen achievement gaps but also impact how children see themselves in relationship to mathematics learning" (National Council of Teachers of Mathematics p. 26). We are all familiar with shifting students into different groups based upon teacher recommendations or test scores. The NCTM gives examples of different ways this occurs in elementary school such as high achieving students being given to one teacher. This practice leads to labels which creates a poor image in students. The book admonishes the need to eradicate ability grouping by providing shared inquiry-based activities and minimizing separated instruction. Project-based learning supports all students by providing appropriate stretch and support because teachers can use the rich context to provide constant support for all students. This brings us to our final point – equitable instruction.

Equitable Instruction

In addition to ending practices such as tracking, NCTM recommends "mathematics instruction should be consistent with researched-informed and equitable teaching practices that *nurture children's positive mathematical identities and strong sense of agency*" (emphasis ours) (National Council of Teachers of Mathematics p. 45). Instruction is not just about students learning mathematics. It is about students seeing themselves as well. Nurturing a child's positive mathematical identity means they need to see people who look like them doing mathematics. We cannot remember being presented with mathematicians of color. Can you think of any? It is difficult for children to believe they can be someone or do something they have not seen done by people who look like them. I, Telannia, attribute my own ability to complete calculus in high school to having a teacher who looked like me – an African American female. It is imperative for teachers to develop students' competency and identities. This can be done by reading books with mathematicians or scientists who are people of color. It is also accomplished when students solve a real-life problem themselves. Again, Project-Based Learning puts students at the center of the situation and allows them to be the solution.

Helping students to see themselves is quite a change. However, *Catalyzing Change*, calls for implementation of practices that promote student agency as well (National Council of Teachers of Mathematics p. 50). This requires students to engage meaningfully with the study of math. The way students engage in the study of mathematics has a profound effect on the learning of mathematics.

> The ways students engage in mathematical modeling in prekindergarten through grade 8 provide a basis for more sophisticated modeling processes. For example, if students have

experience identifying and making assumptions about real-world situations during their first few years in school, they will be aware that solving real problems involves making assumptions when they encounter modeling problems in later grades. Early exposure to ambiguous or open-ended problems can help children become comfortable with the idea that useful, informative solutions to real problems are neither perfect nor unique. Students can develop a mathematical modeling disposition and competence in individual aspects of the modeling process and later combine these aspects into a complete, iterative process.

(Bliss and Galluzzo)

Common Core is an excellent model for how students should engage with learning mathematics that is research-based. The power is in the mathematical practices. As the opening to the Standards for Mathematical Practice states, "The Standards for Mathematical Practice describe varieties of expertise that mathematics educators at all levels should seek to develop in their students. These practices rest on important 'processes and proficiencies' with longstanding importance in mathematics education" (Council of Chief State Officers and National Governors Association for Best Practices). For the first time, an importance on how students engage is put above what students engage in to learn. These are the habits of mind students should have with all mathematical content. All the practices work individually and collectively to create the learning environment described by NCTM's *Catalyzing Change*. Knowing that students need to have a classroom with these practices at the forefront is not the same as knowing how to make these practices be at the forefront. The practices which produce a desired learning environment are all connected to Project-Based Learning which is the "how" for teachers and students. However, CCSS mathematical practice number 4 – model with mathematics – is the greatest connection of them all.

According to CCSS, modeling mathematics involves "students applying the mathematics they know to solve problems arising in *everyday life, society, and the workplace* (emphasis ours)." Modeling can be simple as well as complex.

In early grades, this might be as simple as writing an addition equation to describe a situation. In middle grades, a student might apply proportional reasoning to plan a school event or analyze a problem in the community. By high school, a student might use geometry to solve a design problem or use a function to describe how one quantity of interest depends on another.

(Council of Chief State Officers and National Governors Association for Best Practices)

The document goes on to explain that modeling mathematically involves making assumptions, approximations, simplifying complex situations, reflecting on results and modifying results as necessary. Assumptions, approximations, simplifications, reflections and modifications are all skills that are embedded within Project-Based Learning as well as the other mathematical practices. As students work toward solutions to real-world problems, they will make sense of problems and persevere in finding solutions. They will use reasoning, construct an argument and use tools to ensure precision. All while examining structure and applying reasoning. This is an example of how much the practices work together in a project. Although most of the practices are involved in a project, not all content standards work to be in a project.

The wording in content standards helps you identify the ones that are prime for Project-Based Learning. For instance, CCSS often uses specific wording and notation in the content standards that identify which content should have a real-world focus. Often the standards use the following words: scenarios, real-world or word problems. For example, in kindergarten we see "Represent addition and subtraction with objects, fingers, mental images, drawings, sounds (e.g., claps), acting out situations, or equations." The key word in this standard is "acting out situations." However, you can also imply it from the word object. There are times where "word problems" is actually stated in the language of the standard. In first grade we see

> Use addition and subtraction within 20 to solve word problems
> involving situations of adding to, taking from, putting together,
> taking apart, and comparing, with unknowns in all positions,
> e.g., by using objects, drawings, and equations with a symbol
> for the unknown number to represent the problem.

Determining the standards to make into a project is the hardest part for many educators. Therefore, we include in Chapter 3 of this book those CCSS that are supportive of Project-Based Learning. We know that not all states are using these standards but you should be able to find a complementary standard in your particular state standards.

Although words like real-world or word problems in the standards is an aspect of modeling, modeling mathematics is more than a word problem. Textbook word problems often deter a student from the mathematical practices because they follow a predictable format. Modeling mathematics is more complex. It is the application of mathematics in the context of the *current* world. The word current is emphasized because application of mathematics changes as technology advances. What is done this year may not be done two, three or five years from now. A very common example is retail. Think about the evolution of sales of goods and services to people over time. Many may remember how people had to calculate the cost using paper and pencil. Now, devices that take cards as well as cash are used and the device does all of the calculation. This does not mean that the employee does not need to know how to do operations with numbers. However, quickly seeing if the calculation done by the device is reasonable, is the new goal.

Additional Reasons for PBL

Now let's close this chapter with more reasons why Project-Based Learning is the perfect tool to teach K–2 students' essential mathematical concepts. We hit on some items but we want to talk about how it is a great structure for you as a teacher while addressing some common concerns. We both were introduced to Project-Based Learning when we first started teaching. We were also both not always teachers. We had multiple career experiences before going into teaching. When Project-Based Learning was presented to us, it made so much sense and we dove in headfirst to figure out how to make it work with mathematics being the central focus of the work. However, as we have worked with teachers from around the world, many were hesitant to dive in. Let's address some of the main concerns we have heard.

When you teach kindergarten to second grade, you teach all the subjects for the entire day and you often feel you don't have time to cover all the material you are required to do. This is an understandable concern but projects allow you to compound time rather than lose it. Project-Based Learning involves multiple subjects at a time. Not all subjects have equal showcasing but at least two subjects are working simultaneously. Rather than thinking of individual subjects within a narrow window of time, you will have subjects working together sometimes in the same window of time. In Section 3 of this book, there is an example of a project for every grade level. All of them are tied to language arts standards. The math and language arts time segments of your class work together rather than separately, which compounds your time as well as deepens your students' understanding in a way that is natural to them.

Often in training we help teachers overcome their fear of no time and too much in their curriculum but there is still a hesitancy to do projects with math content. Sometimes it is because teachers are least comfortable with math content. Others can't think of how to teach math deeply in a project. Both are understandable reasons why you would be hesitant. We have provided many chapters in this book to help you overcome these challenges. However, we want to address a few right now to hopefully move you to keep going on to other chapters in this book. If you are uncomfortable with math content, planning a project is the best way to become comfortable. Between the two of us, we have dozens of hours of math coursework and over 40 years of teaching secondary mathematics, yet it was when we did a project that we learned mathematics in a deeper way. Even if you use someone else's project plans, you develop a deeper understanding of your work to adapt it to meet the needs of your students. We have great projects to get your started.

Not being able to think of how to do a project with math as the star content is difficult for so many. Even after years of doing it, we still struggle. Some of the struggle is because you don't know how the math relates to an authentic situation or problem. This can be fixed by just asking family and friends how they use math. This will enable you to start to think of other projects you can do beyond what is provided in this book or on our website mathpbl.com. You can also get better at this by understanding the contexts for math.

Mathematics is the one subject that is normally taught without context and therefore we don't know context beyond immensely basic ones such as counting or simple operations. Some of this is because we have a hard time seeing that the math used in other subjects is still math. For example, reading and writing across the curriculum is standard practice. Everyone understands how these two skills are used no matter what the subject and we wouldn't be comfortable saying we can't do it. However, math is not seen as something that occurs across all the subjects as well. It is a literacy skill just with numbers. Yet, we easily don't look at the math we use in science as still being math. It didn't turn into a science skill. It is a math skill used in science. We like to frame it like this to teachers: math and English are the only two skills a student develops. These skills are applied in different contexts that we sometimes name different categories like social studies, science, art or technology. This is great news given math and English are often the two subjects teachers have to emphasize most. It is impossible to do a project where math or English is not one of the main subjects of focus.

Did we convince you to do Project-Based Learning using math as your focus subject? We sure hope so. Check out another chapter. We suggest experiencing a project which is in Chapter 2.

Reference List

Bliss, Karen, and Ben Galluzzo. *Guidelines for Assessment & Instruction in Mathematical Modeling Education, GAIMME*. Edited by Sol Garfunkel, Society for Industrial and Applied Mathematics, 2016.

Council of Chief State Officers, and National Governors Association for Best Practices. "Common Core State Standards Initiative." *Common Core State Standards.org*, n.d., http://www.corestandards.org/Math/Practice/.

Delpit, Lisa. *Multiplication is for White People: Raising Expectations for Other People's Children.* The New Press, 2012.

Flynn, Mike. *Beyond Answers Exploring Mathematical Practices with Young Children.* Stenhouse Publishers, 2016.

Liljdahl, Peter. *Building Thinking Classrooms in Mathematics.* Corwin Mathematics, 2021.

National Council of Teachers of Mathematics. *Catalyzing Change in Early Childhood.* National Science Teachers Association, 2020.

National Council of Teachers of Mathematics. *Principles and Standards for School Mathematics.* NCTM, 2000.

Youcubed.org. "Fluency Without Fear." *Youcubed.Org,* January 2015, https://www.youcubed.org/evidence/fluency-without-fear/.

CHAPTER 2

Experiencing PBL

A Professional Learning Simulation

This chapter serves as a walkthrough of a first-grade mathematics-based PBL unit. It is designed to help you experience Project-Based Learning and develop an understanding of how other subjects work together with mathematics in a PBL unit. You will experience multiple strategies and structures that are research-based practices used in projects. Resources will accompany strategies and structures mentioned. Given literacy is equal in prominence as mathematics, the experience includes effective literacy strategies. Chapter 4 is dedicated to how literacy aligns with mathematics. The experience is scripted so that you can see the flow from both the teacher and student viewpoints.

To make this as realistic as possible, we utilize a few of the blocks of time from a common daily schedule (see Chapter 3 for full details). The blocks of time are morning meetings, math and literacy. Rather than just reading the chapter, consider completing this simulation with your grade level team, during a faculty meeting or with family/friends. You should primarily experience the process from the perspective of a student. Reflection/discussion questions follow each segment of the simulation to help you reflect on the practices as a teacher. Teachers new to PBL and veteran PBL teachers will grow by walking in the shoes of their students.

Before you begin, note there are handouts referenced throughout the simulation. All the handouts, as well as a link to downloadable versions, can be found in the Appendix in the Experiencing PBL Handouts section and

DOI: 10.4324/9781003237358-3

www.routledge.com/9781646322114. Although the simulation is primarily analog, it can be done with the support of technology or from a distance. Digital methods are referenced during the simulation. Distance learning guidance is provided in graphical text boxes. We recommend you alternate between digital vs analog during the simulation. You should also try the distance learning context.

We are culturally responsive teachers. Zaretta Hammond defines culturally responsive teaching as

> an educator's ability to recognize students' cultural displays of learning and meaning making and respond positively and constructively with teaching moves that use cultural knowledge as a scaffold to connect what the student knows to new concepts and content in order to promote effective information processing. All the while, the educator understands the importance of being in relationship and having a social-emotional connection to the student in order to create a safe space for learning.
>
> (Hammond p. 15)

Throughout the book, she explains how it is a way of being rather than executing strategies from a bag of tricks. We have studied her work and highly recommend her book. Our way of being a culturally responsive teacher is all throughout this book including this simulation. As I, Telannia, worked to create the simulation, I thought about my own students and ways I work with them. My students are diverse. In any given school year, there are at least 30 different languages spoken by our students. Hammond's culturally responsive framework has been vital to my practice.

During the simulation, you may want to take turns, with one person reading the teacher's instructions and completing the teacher's actions. The goal, however, is to experience this simulation from a learner's perspective. Treat it like a classroom experience. Complete typical transitions from one segment to the next just like you do in your classroom. For example, you could say, "If you hear my voice, clap once. If you hear my voice, clap twice" to get the students attention and set up the next portion of the class. Given the possibility of a K–5 school completing this as a book study, the project idea for the simulation is the same in both the K–2 and 3–5 versions of the book. Just the standards and the math simulation are different.

There are a lot of different resources used by schools. To make the simulation as accessible as possible, the simulation was designed using Eureka Math Curriculum (Great Minds) which is free and widely used. The context for the simulation is the following Common Core State Standards which occur in Module 2 of Eureka Math's first grade textbook:

❏ 1.OA.1 Use addition and subtraction within 20 to solve word problems involving situations of adding to, taking from, putting together, taking apart, and comparing, with unknowns in all positions, e.g., by using objects, drawings, and equations with a symbol for the unknown number to represent the problem.

❏ 1.OA.2 Solve word problems that call for addition of three whole numbers whose sum is less than or equal to 20, e.g., by using objects, drawings, and equations with a symbol for the unknown number to represent the problem.

❏ 1.OA.3 Apply properties of operations as strategies to add and subtract. (Students need not use formal terms for these properties.) Examples: If 8 + 3 = 11 is known, then 3 + 8 = 11 is also known. (Commutative property of addition.) To add 2 + 6 + 4, the second two numbers can be added to make a ten, so 2 + 6 + 4 = 2 + 10 = 12. (Associative property of addition.)

❏ 1.OA.4 Understand subtraction as an unknown-addend problem. For example, subtract 10 − 8 by finding the number that makes 10 when added to 8.

❏ 1.OA.6 Add and subtract within 20, demonstrating fluency for addition and subtraction within 10. Use strategies such as counting on; making ten (e.g., 8 + 6 = 8 + 2 + 4 = 10 + 4 = 14); decomposing a number leading to a ten (e.g., 13 - 4 = 13 - 3 - 1 = 10 - 1 = 9); using the relationship between addition and subtraction (e.g., knowing that 8 + 4 = 12, one knows 12 - 8 = 4); and creating equivalent but easier or known sums (e.g., adding 6 + 7 by creating the known equivalent 6 + 6 + 1 = 12 + 1 = 13).

To select appropriate literacy standards for the project, I had to think first about the literacy that is naturally occurring in the project. After I listed all the literacy that occurs, I then looked at the curriculum guide for my school district to see what literacy occurs at the same time as the math module from Eureka Math. There are three units that happen at the same time as the math unit. I selected the literacy standards that most aligned with the list I made. That resulted in the following common core standards to focus on:

❏ W.1.6 With guidance and support from adults, use a variety of digital tools to produce and publish writing, including in collaboration with peers.

❏ RL.1.1 Ask and answer questions about key details in a text.

❏ RL.1.3 Describe characters, settings, and major events in a story, using key details.

The simulation begins on the second day of the project. During the project launch on the previous day, students were introduced to the project and created

Know	Need to Know
• Families are sometimes homeless	• Who do we ask for donations?
• Cats and dogs are sometimes homeless	• Where are we going to put the store?
• Homeless means you don't have a place to stay	• Why are they homeless?
• A person you see holding a sign on the street may be homeless	• Can we help them find a home?
• Mrs. Ramirez works at the shelter	• Who can we ask to donate items to the store?
• We are going to host a shopping day for the people at the shelter	• Can we put toys in the store for the kids?
• We have to ask people to donate items for our shopping day	• What do you need to create a shopping day?

Figure 2.1 Know/Need to Know list

initial questions (see Figure 2.1). Students understand they are going to partner with the local women and families shelter to create a shopping day. They heard from the director of the shelter and understood that the families there rarely get to shop and they would feel good to "shop." They understand they are asking the community for donations, collecting/organizing clothes and hosting a shopping day at the local shelter. Other options for the final product could be the shelter coming to the school or a shopping day for students in the school.

EXPERIENCING PBL SIMULATION
Morning Meeting

Materials

- ❏ Entry event review [Google slideshow] (available on mathpbl.com)
- ❏ A large bulletin board (a wall with paper and border to look like a bulletin board will work)
- ❏ Bulletin board letters to spell out, "How can we, as a retail owner, create the best shopping day?"
- ❏ Centers' sign-up display
- ❏ Duty roster display
- ❏ Poster paper

Classroom Structure Background

- ❏ Morning routines range from setting up activities for the day to class building. These routines are often housed in a "morning meeting." It just depends on the teacher and school goals. During a project, morning meetings become an integral part of the process. Where the morning meeting is conducted, students find key project visuals such as what students know about the project, questions created as a class, the project question and key dates.

Classroom Set-Up

- ❏ **Knowledge of students:** given the nature of this simulation, it is critical to know if anyone has experienced or is experiencing homelessness. When you do the simulation with other adults, ask them to complete an anonymous survey to see if anyone has experienced homelessness. If they have, inform them the context will be homelessness and they are welcome to not participate. Let them know the issue will be handled with dignity and respect. If you choose to do this project with students, you will need to connect with school counselors and admin to see if any students will be impacted and to alter the project or keep the privacy of those students.
- ❏ **Morning meeting set-up:** although this step is optional, it is encouraged for you to set up an area of the room as it would be in the actual project.

There should be a floor mat next to a bulletin board. On the bulletin board, the project question, "How can we, as a retail owner, create the best shopping day?" is written at the top. Underneath the question, are the days of the week along with key deadlines. The final item is posters that contain what the students know and the questions they have (see Figure 2.2). You will also need to display some images that are in a Google slideshow. This slideshow can be downloaded from mathpbl.com.

There are certain routines that occur outside of a project that continue to happen in a project. Some of these routines occur in this simulation. In the room where you can conduct the simulation, you can also have a section in the room that has a mood meter, word wall, centers' sign-up (see Figure 2.3) and duty roster. Mood meter is a board that has different feeling designations such as happy, okay, sad and angry. Students move their name to match their mood. You can have a wall that shows various students' duties for the day as well as sign-up area for centers. These routines occur in the simulation but are not required for the experience. Word walls are typically on the same board as the project information but they can be placed in another area.

❑ **Classroom roles**: in Project-Based Learning, students and teachers share power by the teacher giving the students voice, choice and agency in the classroom. One way of doing this is giving students classroom duties that they do in and out of a project. This is not uncommon in a classroom but it is vital to support some of the even greater work that happens in a project. For this simulation, the duties are from EL Education's video on classroom responsibilities (EL Education). Duties are a familiar practice. If you are new to them, we recommend you watch the video. Before the simulation, assign people the different roles and explain to them what they will do during the simulation. Everyone needs to be in groups of three or four. Each person in a group would be given a role (see Figure 2.4). Here are the responsibilities along with a short explanation:

■ Retriever: gets and returns materials for the group
■ Paper Manager: passes out papers and puts papers in the turn in bin
■ Custodian: cleans off desks after every activity that causes a mess (i.e., breakfast or art)
■ Substitute/Floater: if someone is absent or not in the room, you do their assigned job. They also help anyone as needed.

How can we, as a retail owner, create the best shopping day?

N/N2K List

Know

- Families are sometimes homeless
- Cats and dogs are sometimes homeless
- Homeless means you don't have a place to stay
- A person you see holding a sign on the street may be homeless
- Mrs. Ramirez works at the shelter
- We are going to host a shopping day for the people at the shelter
- We have to ask people to donate items for our shopping day

Need to Know

- Who do we ask for donations?
- Where are we going to put the store?
- Why are they homeless?
- Can we help them find a home?
- Who can we ask to donate item to the store?
- Can we put toys in the store for the kids?
- What do you need to create a shopping day?

Calendar

Mon	Tues	Wed	Thur	Fri
	Understanding & Summing Up			Draft Request
			Donation Request	
				Final Count
		Set up the Store	Shopping Day	

Figure 2.2 Bulletin board

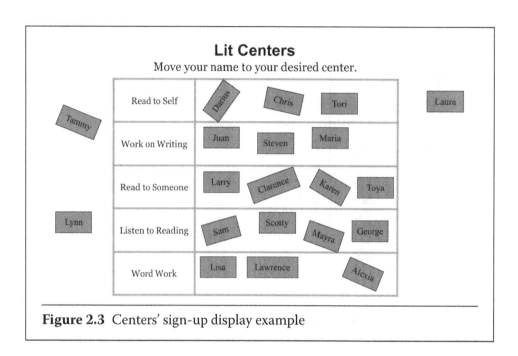

Figure 2.3 Centers' sign-up display example

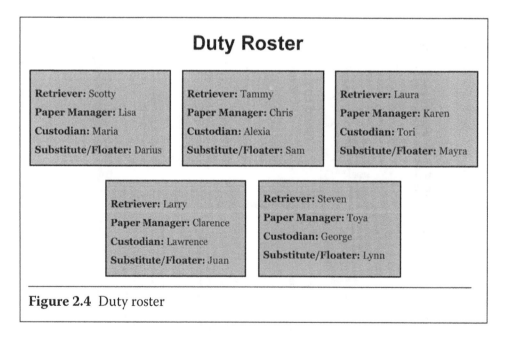

Figure 2.4 Duty roster

DISTANCE LEARNING

Learning management systems are the backbone to distance learning. They house all of the resources for students to access. The project bulletin board changes into a digital form and is on the home page. Bitmoji (https://www.bitmoji.com/) grew in popularity when classrooms moved to distance learning due to Covid-19. This is a great way to create a project board and represent a virtual classroom in general. Catharyn Shelton wrote a great article on how to design a culturally responsive Bitmoji classroom (www.edutopia.org/article/how-design-culturally-responsive-bitmoji -classroom). The board not only represents a look at the visual context of your classroom but it can provide links to needed items.

Bitmoji classrooms allow for students to click on images that take them to a link. You will want to make the "know/need to know" lists, books that are read as a class, handouts for the day, mood meter and virtual sign-ups as links. With the exception of the books and mood meter, all of the links are to Google documents. The books are direct links to electronic books while the mood meter is a digital survey. The duty roster can still be used but changed to reflect the needs of how they work together in a breakout room (recommended for 2nd grade only). One person would remind people to stay on track with the talking (Tracker), one person would call on people to talk (Leader), one person would help keep time (Time Keeper).

To support the delivery of information that would be in a slideshow, it is recommended you use a tool like Nearpod (www.nearpod.com) or Peardeck (www.peardeck.com). These tools provide a great way to show students information while also providing the ability to get reactions from the students.

Sequence

1. From your chair or standing in front of the students as they sit on the morning meeting mat, begin the meeting with the following script:

Student Actions: Students talk in pairs about how they are feeling and then move their name on the mood meter or they fill in the Google form.

Good morning students! Do you remember how when I greeted you this morning, I said, "Are you excited about our thinking today?" You said, "Yes!" I am excited about our thinking as well. Let's begin our morning meeting. As always, let's check in with each other. Turn to another person and tell them how you are feeling right now. Remember, you take turns and you show appreciation for someone sharing. For example, I am really happy today because I am going to the movies tonight. The person who was listening then says thank you so much for sharing that with me. I am so happy for you. When you are done, you go to the feeling board on the wall and place your name under happy, okay, sad or mad (This can be done digitally using a Google form (example at mathpbl.com) where students pick the image that matches them.)

Student/Teacher Actions: Individual students respond to questions while teacher asks the entire class to agree with the answer using thumbs up or thumbs down. If most do not agree, another student is called to answer. Here are the answers to the questions:

❏ Mrs. Rameriz works at the women's shelter
❏ The place is a women's shelter near the school
❏ We are hosting a shopping day for the women and children

2. Take attendance and complete any lunch count requirements while students are talking. Observe the mood meter to make note of who to support that is sad or mad. Quickly talk with them and let them know you are glad for them sharing and that you will help them all day.

3. After you check in with students, go back to your chair or stand in front of the class and say: *Thanks so much everyone for sharing how you are feeling. We are a community which means we care about how each other are doing. We also help our greater community which is people who live around us. Yesterday, we found a need in our community and we agreed to help. Let's talk about what happened yesterday.*

4. Cue up the Entry Event Review slideshow, move to the first slide and say: *Let's see what you remember from yesterday. Who is this person?* After a brief share out, move to the next slide and say: *What is this place?* After a brief share out, move to the next slide and say: *What are we doing for the shelter?* (See student/teacher actions.)

5. Go stand by the list of know and need to know and say: *You guys are so good. Let's keep looking at what we are doing in this project. Look at this list of what we know.* Read the list of items in the know column from the image in the setup section then say: *Is there anything else we know?*

6. Then say: *Now let's look at what we need to know.* Read the list of items in the need to know column from the image in the setup section then say: *Is there anything else we need to know?*

7. Now say: *When we do projects, we always work to answer our questions. Sometimes we are working on one question. Sometimes we are going to try to find the answer to more than one question.* Point to the questions that say, why are they homeless? and What do you need to create a shopping day? *Yesterday, we highlighted these questions as what we want to work on today. These questions connect to what we have to do today.* Point to the spot on the board that says, "Understanding & Summing Up." *Today, we are going to work on understanding the shelter and summing up what we need to have a shopping day. This is what we said we needed to do today. This means we are going to have to put our thinking caps on.* Make a motion of putting a cap on. *We are also going to have to work together. Let's get ready to work together. First, let's pick which learning centers we want to be in for this afternoon. Let's review the literacy centers and then we will go in small groups to go select our center for today.* Move the Google slide to show the literacy center options. Explain each option as follows:

- *Read to Self: this is where you select a book from the class library and you sit in the place of your choice somewhere in the room.*
- *Work on Writing: this is where you get to write about anything you want.*
- *Read to Someone: this is where you read with someone but you have to be with your partner in the reading buddy section of the classroom.*
- *Listen to Reading: this is where you listen to an audio book.*
- *Word Work: this is where you work with words from the word wall. You get to talk with other classmates how the words relate to each other.*

Okay, take a moment to think about which one you want to do and then I am going to have you go move your name to what you want to do today.

Teacher Actions: Call on students who raise their hand. Add their responses to the list. Even add the information if it is not correct. After a few days of doing the project, you will begin to also ask students if there is anything we wrote that is not correct and guide them to changing the wrong answers.

Teacher Actions: Call on students who raise their hand. Add their responses to the list. Remember to say thank you for that question for every question offered.

Teacher Actions: Take a picture of the students choices to keep track of the results. Student's choosing is key to agency, however make sure they don't pick the same thing thing every day.

Teacher Actions: While students complete their role, support them with reminders or modeling. Also use this time to set up for the literacy lesson.

8. Give the students a minute and then start to call groups of students to move their name. If you didn't set up the section for the simulation, just ask each student to share their selection and write the list on a whiteboard or add it in the Google slides. The learning centers will happen later in the day and are not a part of this simulation experience. However, we have found it is a good practice to set up during the morning meeting.

9. Then say: *Now let's look at our math centers. Just like the literacy centers, you have a choice. Let's review the math centers and then we will go in small groups to go select our center for today.* Move the Google slide to show the math center options. Explain each option as follows:
   - *Math Play: this is where you play a math game by yourself or with other classmates.*
   - *Twist and Turn: this is where you use manipulatives, remember manipulatives are stuff like blocks or cut outs, to solve a problem.*
   - *Journaling: this is where you get to write about what you just learned.*
   - *Picture Perfect: this is where you explain the math through pictures.*
   - *Up Your Game: this is where you work with classmates to work on a problem that is connected to what we are learning today.*

10. Then say: *Okay class, we are almost ready to start our day. Now it is time for us to get our materials set up for the day. Check out the duty roster and those who are retrievers and paper managers go quietly to get the materials for your group. Remember the materials are in that corner of the room. You have three minutes to do your duty and then return to the mat. Your three minutes start now.*

11. Teacher opens the electronic book *June Peters, You Will Change the World One Day* by Alika Turner (Turner) and projects it on the screen or pulls out the paper copy. Display near the mat a poster paper that is a bubble map with June in the center circle. It must be in a place where you can write on the poster. Near the mat but not covering the bubble map is a tree map of how good readers ask questions (see Figure 2.5).

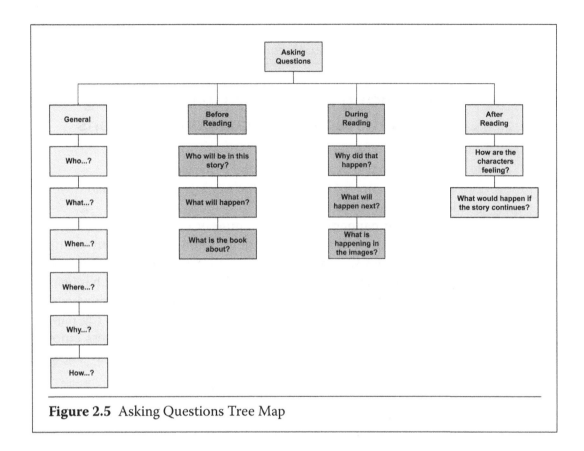

Figure 2.5 Asking Questions Tree Map

Experiencing PBL Study Guide: Morning Meeting Reflection/Discussion Questions

1. How did you feel as a student in the experience?
2. What were some of the things that occurred that you had never done as a student before?
3. This is what happens at the beginning of the day of a project. It is a combination of an entry event and how to constantly set up the learning for the day. Although an Entry Event happens in one day for upper level elementary, for K–2 the entry event happens for a few days. An entry event is designed to capture your heart before asking your head to follow. The huge emotional tug happens on the first day but the subsequent days can be less emotionally focused. How did the review of the previous day tug at your heart?
4. The routines in this morning meeting had some things that may have been familiar and unfamiliar. How was the morning meeting similar and different from your current practice?
5. Skim the teacher's actions during the morning meeting. What stood out to you? What is the purpose of reviewing the know/need to know lists? Why is it important to let students have voice and choice in the learning centers? What do you have questions about?
6. Given being a culturally responsive teacher is a way of being, Zaretta Hammond (*Culturally Responsive Teaching and the Brain*) created a framework that has key areas for teacher capacity building. The areas are Awareness (of culture), Learning Partnerships (relationship between student and teacher), Information Processing (challenging work for students) and Community of Learners and Environment (student agency and safety). How did you experience this framework in this experience?

LITERACY BLOCK

Materials

- ❏ *June Peters, You Will Change the World One Day* by Alika Turner projected on a screen or printed out.
- ❏ Bubble Map Anchor Chart with June in the center circle
- ❏ Tree Map Anchor Chart of Asking Questions
- ❏ Handout:
  - ■ Bubble Map for Character Description

Classroom Structure Background

- ❏ **Literacy block:** this typically occurs before lunch and close to the math block. Everyone does something different in this block of time. For example, some teachers follow a program purchased by their school or district like Benchmark Literacy. The block of time ranges from 45 minutes to 90 minutes. Typically, the time is determined by the school or district. In projects, the content is taught through the lens of the project. School materials can be used but they are adapted to support the project. It is also common for this block to work together with other subject blocks to the point where they are completely blended together.

Classroom Set-Up

- ❏ **Literacy block set-up:** this set-up is essential for the experience. The book, *June Peters, You Will Change the World One Day* by Alika Turner, should be printed or projected on the screen. The project wall with the same information as in the morning meeting will be set up. Near the project wall or on it should be a word wall section. On a card should be the word *character*. On another card should be the definition *is a person in a book*. Tape the cards together so that the definition folds behind the word. Hide the definition on the word wall.

On a whiteboard, write the objectives and agenda for the literacy block. The objectives are: I can explain why a person is homeless; I can describe a character in a book; I can ask questions while I read. The agenda is: describe June as a class; read with a partner and describe a character; share character with the class. Hang the ask questions anchor chart (see Figure 2.6) where it is visible. Also hang chart paper that you can write on that has the name *June* in the middle of a circle in the center of the paper.

On each student desk is a sheet of paper with the *word character* at the top and an open circle in the center. In a section in the room is a variety of books about homelessness for people to pick from. There needs to be at least one book for every two students. If possible, audio books should be available. If you do not have a lot of library resources that involve the topic of homelessness or the ability to print free book resources, you can have other books and still complete the experience.

 DISTANCE LEARNING

There are a few ways you can execute the reading of the book using virtual conference tools. Some require a moderate level of comfort with technology. The first way requires you to be in a classroom and students can watch you as if they were in the classroom with you. You will need to have a printed copy of the book. Another option is to have a link to the book in the virtual classroom. Students would read along with you. You would still demonstrate the creation of the anchor chart with students watching you on video. The final way is to incorporate the tool into Nearpod (www.nearpod.com). This tool allows you to annotate on a slide so that students can see it as you write on it. It can also allow you to have students interact with you by drawing their own observations.

Just like in the morning meeting, you can execute many of the other pieces such as reading a word from the word wall in your classroom with students watching. You can also have students continue to work in the Nearpod for the experience. The handout of their own character description could be done in Nearpod or on a copy of a Jamboard that they click in the Bitmoji classroom.

For the reading portion, students can choose books by clicking on a book from the library in the Bitmoji classroom section. In order for students to read in pairs, you will want to put them in breakout rooms together. (Recommend breakout rooms for 2nd grade only.) Remember to check on each group every few minutes.

Sequence

1. Stand by the list of questions students generated. Tell students: *Good job class! Thank you so much for getting us ready to learn. Let's look again at one of the questions we are working on.* Point to the question – "Why are they homeless?" *Can someone read the question for me?* Call on a student and point to each word as they read it. *Thank you so much. When we have questions, it is helpful for us to think about what it means for us to be able to answer the question. It helps focus our minds on what we need to do. All you have to do is change the question into a sentence. Let's look at the sentence I made for this question.*

2. Move to where the objectives are written. *Let's read this sentence together.* As you and the class say the objective, point to each word. *I can explain why a person is homeless. I put the words "I can explain" before part of the question and now it is a sentence that helps me know what to do. I have to explain to answer this question. Anyone have any ideas of how we can find out the answer?* Take some suggestions from the class. Possibly remind them of ways we found out answers to questions before if they have trouble thinking of answers. *Thank you everyone for those ideas. I appreciate it so much when you share your thoughts.*

3. Move to the word wall and pull off the word character. *Today, we are going to read to understand why people are homeless. This will require us to know how to describe characters. Can anyone tell me what the word character means?* Call on a few students. Thank them for sharing. Reveal the definition of the word. *Character is a person in a book.* Place the word back on the word wall with the meaning revealed instead of hidden.

> **Student Actions:** Students are being led through a mini-lesson. They are sitting on a mat following the teacher's directions.

4. Move back to the list of objectives. *This brings us to our other objectives for today. Let's read the sentences together.* Point to each word again as the class reads. *I can describe a character in a book. I can ask questions while I read.*

5. Move to the agenda board. Point to the morning literacy agenda and read the list. *We are going to describe a character named June. After we describe as a class, you are going to read with a partner and describe a character from your chosen book. We will come back together to share our character with the class.*

6. Move to the "Ask Questions" anchor chart. *Before we read, let's review one of the things good readers do. As the objective said, you are able to*

ask questions while you read. Here are some common questions. Let's answer the questions under "Before Reading" together.

7. Show the cover of the book or point to the cover on the projector screen. Say, *Look at the cover of the book. Who will be in this story?* Take a few answers. One should be a person named June. Try to see if they can predict more based upon the part in the title Change the World One Day. Remember to thank everyone for their answers. *What will happen?* Take a few answers. *What is the book about?* Take a few answers.

8. Say, *Before we read and see if we are right, let's look at how we describe a character in a book.* Write at the top of the anchor chart with the circle for June the word character. *Pictures and words help us to describe a character. As we read and look at pictures, we pay attention to how the people look, act, think and feel.* Write the last part of the statement on the anchor chart. See Figure 2.6.

9. Point to the cover of the book. Say, *Based upon this picture of June on the cover. Can someone tell me one word to describe June?* Pick on a student and write what they share in a bubble and connect it with a line to June's name.

10. Turn to the first page of the book so that the words and the image on the following page shows on the screen or in your hand. Read page 4. Model

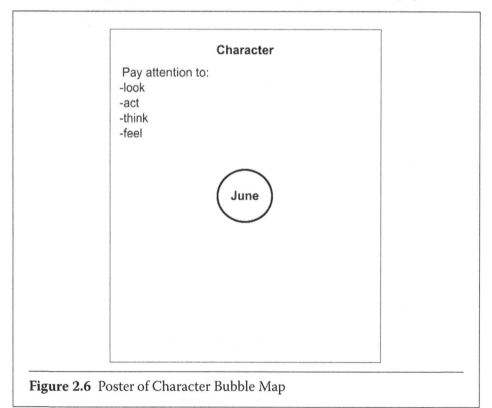

Figure 2.6 Poster of Character Bubble Map

out loud how you find descriptions of June based upon page 4 and 5 of the book. Write some of the descriptions you think are important on the chart.

11. Have students read page 6 with you and look at page 7. Have various students share a few descriptions based upon the reading and image.

12. Say, *It is now time for you to work on describing the main character with a book of your choice. There are several books in the book corner about homelessness. Remember, you must read with a person in your group. Pick a book together and find a comfortable place in the room to read. Fill out the bubble map about your main character in the book. You are reading and describing for the next 30 minutes.*

13. In pairs, students select a book and grab the bubble map that was left at their desk by their classmates during the morning meeting. Students read and describe with their partner.

14. Teacher uses the signal of chimes to get students' attention. Students stop what they are doing and listen for directions. Teacher tells them to return their book. They keep their bubble map and return to the mat.

15. Once students are on the mat, say *It is time for us to share our character descriptions.* Call up the first student pairs to come up next to you to share their character description. The student pairs share their description. *Thank you so much for sharing.* Place the bubble map where everyone can see.

> **Student/Teacher Actions:** Students work in pairs reading their chosen book and writing descriptions about their character for 30 minutes. Teacher confers with each pair. Teacher reminds students of asking questions while they read and how to find descriptions of their character. Teacher identifies three student pairs to share during the debrief portion of the class.

16. Repeat with the next two groups. Remember to thank each one after sharing and put the bubble map where everyone can see.

17. Have students talk first in pairs and then with the whole class about the following questions:
    - What descriptions are the same?
    - What descriptions are different?
    - What do we now know about homeless people?

> **Student Actions:** Students share first with a partner the answer to the questions. Students who are called upon share their discussion with their partner.

18. Transition to the math block of the day. Say, *You have done a lot of work. We are getting closer and closer to helping the families in the shelter. We are going to keep looking at the question of "Why are they homeless?" but we are also going to look at "What do you need to create*

a shopping day?" Point to the second focus question. *Let's move to our desks so that we can work on the second question.*

19. Students move to their desks.
20. Teacher prepares for the math block. Teacher opens up Math Lesson Google slides.

Experiencing PBL Study Guide: Literacy Reflection/Discussion Questions

1. How did you feel as a student in the experience?
2. What were some of the things that occurred that you had never done as a student before?
3. This is the first day that students start to learn literacy skills while doing the project. How was it similar to students learning about a reading skill? How was it different?
4. How did the teacher and students work together during the literacy block?
5. What teaching strategies occurred during the literacy block?
6. What strategies would you want to incorporate in your classroom?
7. How did the lesson support all students, including the gifted and talented, English Language Learners, and those with learning disabilities?
8. Given being a culturally responsive teacher is a way of being, Zaretta Hammond (*Culturally Responsive Teaching and the Brain*) created a framework that has key areas for teacher capacity building. The areas are Awareness (of culture), Learning Partnerships (relationship between student and teacher), Information Processing (challenging work for students) and Community of Learners and Environment (student agency and safety). How did you experience this framework in this experience?

MATH BLOCK

Materials

- ❑ Whiteboard for each student along with a marker
- ❑ Linking cubes of two different colors. A set of 20 for each student
- ❑ Math lesson [Google slideshow] (available on mathpbl.com)
- ❑ Projector or interactive display
- ❑ Handouts:
  - ■ Problem Set (see appendix or www.routledge.com/9781646322114)
  - ■ Exit Ticket (see appendix or www.routledge.com/9781646322114)

Classroom Structure Background

- ❑ **Math block:** this typically occurs before lunch and close to the literacy block. Everyone does something different in this block of time. For example, some teachers follow a program purchased by their school or district like *Everyday Mathematics* by McGraw-Hill. The block of time ranges from 45 minutes to 90 minutes. Typically, the time is determined by the school or district. Just like the literacy block, in projects, the content is taught through the lens of the project. School materials can be used but it is adapted to support the project. It is also common for this block to work together with other subject blocks to the point where they are completely blended.

  Many schools make teachers follow the given curriculum closely or require them to implement it exactly. As a result, this experience followed Lesson 3 in Module 2 for Grade 1 of Eureka Math lesson structure i.e., fluency practice and application problem. (NY) However, the details in the structure are changed to fit the project.

Classroom Set-Up

- ❑ **Math block set-up:** this set-up is essential for the experience. Make sure the math lesson Google slide show is projected. Each student must have a whiteboard at their desk along with a marker. Each student must have 20 linking cubes where ten are of one color and the other ten are another color. If you do not have linking cubes, you can use anything that is a solid color and links together such as Legos. Each student should have a problem set and exit ticket handouts.

 DISTANCE LEARNING

The best way to support this lesson with all of the interaction needed is to use a tool like Nearpod (www.nearpod.com) or Peardeck (www.peardeck .com). It will allow for you to see the students' work and easily access digital linking tools (www.didax.com/math/virtual-manipulatives.html). An alternative is to have students show you their work on paper to the screen as well as for them to use Legos if they have them in their house to represent the linking cubes.

Just like the other simulations, the handouts could be obtained by clicking the link on the Bitmoji classroom board. Recommend using a Google Jamboard if your students are accessing the class via an iPad. If they are using a computer, Google docs would be easier to show their work. You can get an example of a digital student math handout on mathpbl.com.

Sequence

1. Get everyone's attention by using the chimes. Wait for students to direct their attention to you. Repeat the process if students don't direct their attention to you within a few seconds of hearing the chimes. This helps develop the routine.

2. Move the slide to the focus question from the need to know list. Say, *This is our other question to focus on today: What do you need to create a shopping day? Remember when we have questions, it is helpful for us to think about what it means for us to be able to answer the question. It helps focus our minds on what we need to do. All you have to do is change the question into a sentence. Let's look at the sentence I made for this question.*

3. Move to slide with the focus question and the objective. *Let's read this sentence together.* As you and the class say the objective, point to each word. *I can make a list of what we need to create a shopping day. I put the words "I can make a list" before part of the question and now it is a sentence that helps me know what to do. I have to make a list to answer this question. Anyone have any ideas of how we can find out what needs to*

be on the list? Take some suggestions from the class. *Thank you everyone for those ideas. Thinking of our own experience shopping is a great way to start thinking of the list. We will work on creating a list in a little bit. Talking with a person who works at a store is another great idea. I know a person. I will ask them if they can come in and talk to us. What is one thing we know we need for families to shop?* Call on a student. *Yes, clothes. We will need lots of clothes. We are going to need to count those clothes.*

4. Move to the next slide and say, *This brings us to our other objective. I can add up the donated clothes. This is going to require us to get really good with adding up numbers. We have been working on making 10. We are going to continue to work on this but now with the context of donated clothes. Let's begin by warming up our brains.*

5. Move to the next slide. With excitement say, *let's take out 1. We have done this for a few days now. Let's see if you remember what to do.* Move to the next slide where the number is 6 is shown. *Students shout out the answer.* Share how 5 and 1 is the answer. *Let's do it two more times.* Move to the next slide where the number 2 is shown. *Students shout out the answer.* Share how 1 and 1 is the answer. *Last one.* Move to the next slide where the number is 10. *Students shout out this answer.* Share how 1 and 9 is the answer.

6. Move the slide to the "Break Apart 10" activity. *Now, let's break apart the number 10. Write the number 10 on your whiteboard and two lines below just like on the screen. I am going to show a number. Write the number that would make it add up to 10.* Give students a little time to write. Move the slide once everyone is ready.

7. Move the slide several times showing different numbers that are less than 10 so that students can write the two numbers that make 10. The answer to the number displayed is on the following slide.

8. Say, *Let's do one more activity. Let's add 10 first. I am going to show numbers to add. We are going to put together the number that makes 10 first and then continue to add. Write the answer on your whiteboards.*

9. Move the slide several times showing different numbers to add. At first it is two numbers and then it moves to three numbers. Each time the answer is after the numbers to add.

10. Say, *Now that our brains are warmed up, let's work on a problem.* Move to the next slide. *Let's read this problem together.* Point to each word as you say them. *Alexa gave 5 shirts. Her sister gave 9 shirts. A friend gave her enough shirts so that she had a total of 15. How many shirts did her friend give her? Use a drawing, a number*

Student/Teacher Actions: For about five minutes, students write on the whiteboard the two numbers that make 10. The teacher moves through the slides which have the numbers students are using along with the answer on the following slide.

sentence and a statement. Extension: How many more would she need to have 19 shirts?

11. Give them five minutes to work on the problem using their whiteboards. Walk around and select different students to share out how they figured out the problem.

12. Gather a copy of the handout entitled Counting Our Shopping Items, the 20 linking cubes and a pencil. Make sure it is near where the Google slides are projected so that you can get to it easily.

13. Move to the next slide and say, *It is time for us to build on what we know. Look at the paper that says Counting Our Shopping Items. It has pants at the top. Also make sure you have 10 red and 10 green linking cubes. Finally, you will need a pencil. Give a thumbs up when you have those items in front of you.* Give students a moment to gather the items which should be at their desks. Show each item as you state each one. *Remember how I say we always use what we know to learn new information. You are doing a great job of that today. Let's keep it going now. I am going to read the problem. Please read silently along with me.* Read the problem. *What is the problem asking us to do?*

14. Call on various students. Thank them for sharing. Confirm for students that we are finding out the total number of pants. Write on the google slide if you can annotate it or write on a nearby whiteboard the following: _____ + _____. Say, *we have been working on writing an expression for a problem. On the count of three, let's share at the same time what goes in the blanks. One, two, three.* Students share the answer of 9 and 7 or 7 and 9. *Good job. Let's complete that part of the handout.* Move the slide to show where students should write the number 7 on the handout.

15. Show the green and red linking cubes. Say, *hold up your green and red linking cubes. With the green cubes, show the number of pants for Kensie. Use the red for Steve. Hold up the results when you are done.* Look around the room and make sure students have the right number of cubes. *Good job. You have 9 cubes for Kensie and 7 for Steve. Okay, now solve the problem and write the amount on your whiteboard. Hold up the answer when you are done.* Give students time to work out the problem. Look around the room for the answer of 16.

16. Move the slide to show the answer along with the question to ask the class. Have students complete the portion of the handout that is shown. Call on specific students to share how they found the answer using something like a jar with student names on a stick. Students share how they did it. *Thank you so much for sharing.*

17. Move the slide again to a new question (Using the linking cubes, is there a way to make ten with what we have?). Read the question and have students talk with a partner. Move around the room and listen to student conversations. Find a student who moved one cube from the 7 and add it to the 9. Have them share out after students have time to figure it out.

18. Say, *thank you so much for sharing. Let's look at it visually on our paper. We have all of Kensie's pants but then we moved one from Steve to Kensie.* Draw a circle around Kensie and one around Steve's. You can also move to the next slide which shows the circled area. *Oh, my goodness. We now have a new sentence.* Draw a name from the jar to call on. Say, *can you tell us the new sentence?* Let the student answer. *Thank you so much.* Let's see the answer.

19. Move the slide to reveal the answer. Give students a moment to update their paper. Walk around for a moment to make sure everyone has their paper updated. Encourage classmates to help those who may be missing information.

20. Move to the next slide. Say, *class I have another question for you. Say what you think at the same time. Do we have a different amount of linking cubes?* Hold up the 10 and 6 linking cubes we made. The class should say no. *You are so right. So that means we have two sentences that are the same. We started with 9 + 7. What is the other sentence that we created that is the same?* Call on someone from pulling their name from the jar. If they get it wrong, say they are close and hold up the new linking cubes to help. Move the slide to show the answer. Give students time to complete the handout.

21. Move the slide and say *you guys did a great job connecting what you know to something new. You don't have to count on to find a sum. You can make 10 and then find the total. Work with a person at your table on the rest of the paper. Make sure you draw a picture before you work out the problem.*

22. Students work with a partner to complete the remaining problems. Teacher circulates to check on students and select some to share their answers in the debrief. You want students who show understanding of the concept of making 10 but also students who show a common misunderstanding.

> **Teacher Actions:** Give students at least 15 minutes to work on the problems.

23. Use the chimes to call students' attention back. Have the students share their results. With each student group, ask them the following questions:

■ What are your number sentences?

■ How did you come up with your number sentences?

24. After the students share their work. Have a whole class discussion about two specific questions: When you made ten, what do you notice about the addend you broke apart? and What new strategy did we use? How is it more efficient than counting to add on? How can this help us add donated clothes?

25. Have students complete the exit ticket. Have duty students turn in the exit ticket for their group.

Experiencing PBL Study Guide: Math Reflection/Discussion Questions

1. How did you feel as a student in the experience?

2. What were some of the things that occurred that you had never done as a student before?

3. This is the first day that students start to learn math skills while doing the project. How was it similar to students learning within a Eureka Math lesson? How was it different?

4. This was the first time students worked on part of the product. How was working on the product combined with learning the mathematical concepts?

5. How was conceptual development combined with procedural fluency in the handout?

6. How did the teacher and students work together during the math block?

7. What teaching strategies occurred during the math block?

8. What strategies would you want to incorporate in your classroom?

9. How did the lesson support all students, including the gifted and talented, English language learners, and those with learning disabilities?

10. Given being a culturally responsive teacher is a way of being, Zaretta Hammond (*Culturally Responsive Teaching and the Brain*) created a framework that has key areas for teacher capacity building. The areas are Awareness (of culture), Learning Partnerships (relationship between student and teacher), Information Processing (challenging work for students) and Community of Learners and Environment (student agency and safety). How did you experience this framework in this experience?

References

EL Education. "Classroom Responsibilities." *EL Education Classroom Responsibilities*, n.d., https://eleducation.org/resources/classroom-responsibilities.

Engage ny. "Grade 1 Mathematics." *Grade 1 Mathematics Module 2 Topic Lesson 3*, n.d., https://www.engageny.org/resource/grade-1-mathematics-module-2-topic-lesson-3.

Great Minds. "Knowledge on the Go Lessons." *GreatMinds.org*, n.d., https://gm.greatminds.org/knowledgeonthego/knowledge-for-grade-1.

Hammond, Zaretta. *Culturally Responsive Teaching & The Brain*. Corwin, 2015.

Thinking Maps. "Thinking Maps." *Thinking Maps.com*, n.d., https://www.thinkingmaps.com/why-thinking-maps-2/.

Turner, Alika. "June Peters, You Will Change the World One Day." *Freekidsbooks.org*, n.d., http://freekidsbooks.org/.

DESIGNING INQUIRY-BASED TASKS AND PBL UNITS

CHAPTER 3

Creating an Inquiry-Based Task or PBL Unit

Creating a Project-Based Learning unit with math at the center has unique considerations especially in the realm of inquiry. We have spent years learning new ways to fill our classrooms with inquiry. Our own challenges are why we included examples of tasks and PBL units in the final section of this book. We felt this would help all teachers by allowing them to use it to start as well as help them plan their own. This chapter explains the process that we take in designing a unit or a task. It is a result of learning from leaders in PBL as well as our own implementation of the practice. Although this chapter explains the process in steps, you do not have to go in the particular order described. It is just in the order that has commonly happened for us. It is a process we do for creating tasks or PBL units. Because we use the word "tasks" in other contexts, in all of the steps we will use project or PBL as the primary language description. Within each step you will see details about planning of a complete elementary school Project-Based Learning unit that you will see in Chapter 6.

Before we begin, we need to explore the daily routine in a PBL/elementary setting (see Table 3.1). Through exploration we have discovered that although schools can differ widely, most schools have a version of the schedule on the left-hand side of Table 3.2. To create the schedule on the right-hand side, we consulted several elementary teachers who implement Project-Based Learning in their classroom. We have provided two possible schedules for both the "traditional" day and for a "PBL unit" day. You will notice that the morning routine

DOI: 10.4324/9781003237358-5

TABLE 3.1
Example elementary PBL schedule

Possible PBL Elementary Schedule

A. **Morning meeting**
   [See <u>project launch day</u> at end]
   a. Daily routine (date, weather, events, etc)
   b. Explore today's work for project
   c. Look at knows and need to knows
   d. Refine "today's work" based on any new questions.
   e. Journal entry: work plan for the day (compare with entry from the end of the last class)

B. **Content area that needs the most time to complete "today's work"**
   a. Include intervention time
   b. Include work to meet pacing requirements
   c. Update project work related to this content (if needed)

C. **Phonics, reader's workshop, writer's workshop or other ELA-related work not included in section B**
   a. Include intervention time
   b. Add work that might help project progress.

D. **Science or social studies that wasn't included in section B**

E. **Morning eeting**
   a. What have we done today?
   b. What other work do we need to accomplish? (refer to journal entry)
   c. Meet in Groups and assign work for after lunch (teacher assigned or student led)

F. **LUNCH/RECESS**

G. **Math (that wasn't included in section B)**
   a. Include intervention time
   b. Update project work related to this content (if needed)

H. **Phonics, reader's workshop, writer's workshop**
   a. Intervention time (if available)
   b. Tie into the project, if possible.

I. **Project time**
   a. Wrap up work for the day.
   b. Journal entry: work plan for next class

J. **End of day routine** (clean up, reflection, announcements, etc.)

K. **PROJECT LAUNCH DAY:**
   a. **Morning time**
      i. Daily routine (sate, weather, events, etc.)
      ii. Conduct project launch
      iii. Create initial K/NTKs
      iv. Assign groups (optional and can move to Day 2)
      v. Discussion: "What are our first steps?"
      vi. Journal entry: reflection on initial thoughts on the project
      vii. *Special Note:* often the project launch lasts for a few days given kindergarten and first grade students often need reminding of the situation and a renewal of their excitement.
   b. **Content area that drives the project**
      i. Include intervention time
      ii. Include work to meet pacing requirements
      iii. Discussion: what skills will we need to be able to do to be successful in this project?
          [Repeat for each content area – let students discover if a content area is needed or not]
   c. **REST OF DAY AS USUAL**

TABLE 3.2
Traditional vs elementary schedule

	Traditional elementary schedules				Possible PBL elementary schedule		
Time:	School 1	Time:	School 2	Time:	School 1	Time:	School 2
8:50–9:15	Morning work, attendance/ lunch count	7:30–8:05	Arrival/ morning work	8:50–9:15	Attendance/ lunch count	7:30–8:05	Arrival/attendance/ lunch count
9:20–9:35	Morning meeting	8:05–8:30	Breakfast	9:20–9:35	Morning meeting	8:05–8:30	Breakfast
9:35–9:55	Number corner/brain booster	8:30–9:00	Morning meeting/ social studies	9:35–9:55	Literacy connected to PBL	8:30–9:00	Morning meeting
10:05–10:15	Phonics	9:00–10:00	PE/reading remediation groups	10:05–10:15	Phonics connected to PBL	9:00–10:00	PE/reading remediation groups
10:20–11:00	Centers	10:00–11:30	Literacy block	10:20–11:00	Centers connected to PBL	10:00–11:30	Literacy connected to PBL
11:05–11:50	Math	11:30–12:20	Math	11:05–11:50	Math connected to PBL	11:30–12:20	Math connected to PBL and morning meeting
11:53–12:03	Read aloud/ lunch prep	12:20–1:00	Lunch	11:53–12:03	Morning meeting	12:20–1:00	Lunch
12:05–12:55	Lunch and recess	1:00–1:40	Centers/small groups	12:05–12:55	Lunch and recess	1:00–1:40	Centers/small groups connected to PBL

(Continued)

45

TABLE 3.2
(Continued)

	Traditional elementary schedules				Possible PBL elementary schedule		
Time:	School 1	Time:	School 2	Time:	School 1	Time:	School 2
1:00–1:40	Specials	1:40–2:10	Recess	1:00–1:40	Specials	1:40–2:10	Recess
1:45–2:15	Writing/grammar	2:20–2:55	Reading/math pullouts	1:45–2:15	Writing/grammar connected to PBL	2:20–2:55	Reading/math pullouts connected to PBL
2:20–2:35	Reading lesson	2:55–3:00	Prepare for dismissal	2:20–2:35	Reading lesson connected to PBL	2:55–3:00	Prepare for dismissal
2:40–3:10	Science or social studies			2:40–3:10	Science or social studies (could be connected to PBL)		
3:15–3:20	Pack up			3:15–3:20	Pack up		
3:20–3:35	Snack time			3:20–3:35	Snack time		
3:40–3:55	Clean Up/dismissal			3:40–3:55	Clean up/dismissal		

and getting the day started are virtually identical. Where it differs is that the content focus in the morning is based upon what work needs to be completed for the project work time in the afternoon. For example, if the students will be doing a lot of measuring, then the morning might focus on how to use a ruler. Another difference is the addition of a meeting prior to breaking for lunch. This gives each group a chance to reflect on what they have done that morning and what still needs to be done in the afternoon. This schedule is designed to have most of the day be based on the project. You can do a project within the blocks of time connected to the subjects for that project.

Step 1: Search through Your Standards or Investigate your World

As we have trained teachers all over the world, a common question is "How do you get started?" The answer is both simple and complex. You use existing projects and real-world situations or you start with the standards. Using the situations of your students' lives is a common way to start. For example, you see on the news an upcoming fundraiser for childhood cancer. Students have a classmate who is battling cancer. You have students help raise money for the fundraiser. The money raised becomes the tool to help students develop their number sense. Problems to use in math class are all around us once we recognize the inquiry process and the steps in the project flow.

Until you start to see projects in the world around you, you can start with the internet and search for project ideas. A Google search using the terms "Math Project Elementary School," for example, yields thousands of hits. In addition to open searching, there are websites such as Mashupmath.com, Weareteachers.com, PBLWorks.com, Edutopia.com and, of course, social media like Instagram or TikTok. Project ideas are everywhere online. As you become a more seasoned teacher you will see that most (probably close to 100%) of these ideas are for end-of-unit, end-of-semester or end-of-year "projects." If you look at the tasks, they are set up as ways for students to demonstrate their knowledge that they have learned during a set period of time.

It is important to state, right away, there is no problem with seeking out these "projects" as idea generators. The problem comes when you try to execute it just as it is stated. You must break them down into the standards you want to cover and then complete the rest of the steps in our process. You need to make them your own. It is common to just take something and use it as it is from a site. This is not something you can do with projects. Don't skip the step of getting clear on what standards you want to teach. Basically, ask yourself what you want your students to know and be able to do. Put them in "I can" statements to help you have a clear picture.

Rather than start with an idea, you can begin with your standards. If you choose to start with standards, you must know that not all standards lend themselves to a task or a PBL unit. See Table 3.3 for common standards to use in a project. Teachers are always surprised when we say that to them. For example, one of the standards in the Common Core State Standards (CCSS) for kindergarten is "Fluently add and subtract within 5." This standard is not worthy of a task or PBL unit because it doesn't require much inquiry. However, it is a standard that can be connected to a bigger idea which might be worthy of inquiry.

When creating a task or PBL unit, it is often helpful to use power standards. "Power Standards" are the most essential knowledge for student success at every grade level (Ainsworth). These are standards that take up more time in the curriculum. Many districts as well as states have identified the power standards for their teachers. Sometimes these standards are called essential standards rather than power standards. Another way to identify these standards is by the amount of time your district allocates to specific standards. Power standards are the ones given the most amount of time normally.

If your district and state have not identified the power standards, it is a really simple process that should be done with colleagues of the same grade level or subject. It is a great process to do even if your district has identified them. Ainsworth gives a simple set of criteria for identifying the standards: *endurance, leverage, readiness for the next level of learning and for a state test.* The criteria can be thought of as a simple question: "Which standards will students still use as adults (endurance), use in other classes (leverage), are needed for the next grade level (readiness), or will be a major portion of an upcoming state test?" (Ainsworth p. 31) Once you select the standards based upon this question, cluster them together as appropriate.

You must look at the standards for math and other subjects. Projects are rarely single subject endeavors. A common way I (Telannia) help people to see how subjects work in a project is to use the illustration of driving a car. A car can only have one driver. In a project, this subject drives the project and for us this is math. Now, many times there is a person riding besides the driver. This person often helps the driver significantly with navigation or even alerting danger. In a project, this is the second subject which gives a lot of support to the project. However, two subjects is the largest amount you can have in a project that still produces deep learning. Other subjects can occur but they are similar to additional passengers in a car. They are along for the ride but they don't really help the learning. Typically, these are ideas already known and just reinforced throughout the project.

After you have selected the math standards, it is important to look at what other subjects complement the math standards. This could be accomplished by

TABLE 3.3
Common Core Standards appropriate for PBL

Common Core State Standards domain	Kindergarten	First grade	Second grade
Counting and cardinality	❑ Count to 100 ❑ Write to 20 ❑ Understand "How Many?" ❑ Quantities greater or less than another		
Operations and algebraic thinking	❑ Represent addition or subtraction ❑ Number pairs that add to 10	❑ Fluently add/subtract to 20 ❑ Solve word problems with numbers to 20 ❑ Apply properties to add/subtract ❑ Determine unknown numbers in an equation	❑ Solve one and two Step word problems with numbers to 100 ❑ Determine if a group has odd or even members ❑ Add objects arranged in rectangular arrays
Numbers and operations in base 10	❑ Compose/decompose numbers 11–19 as "10 + #"	❑ Count to 120 ❑ Understand <,=,> ❑ Add to 100	❑ Count and skip count to 1000 ❑ Understand <,=,> ❑ Understand properties of addition
Measurement and Data	❑ Describe measurable attributes ❑ Compare two objects by their measurable attributes	❑ Order objects by length ❑ Tell and write time ❑ Organize, represent and interpret data	❑ Measure length Using appropriate tools ❑ Solve word problems with time and/or money ❑ Represent measurement data ❑ Use pictographs and bar graphs to represent data

(Continued)

TABLE 3.3
(Continued)

Common Core State Standards domain	Kindergarten	First grade	Second grade
Geometry	❑ Correctly name shapes ❑ Analyze and compare two shapes ❑ Build shapes from component parts	❑ Understand defining and undefining attributes of shapes ❑ Partition shapes into equal shares	❑ Partition rectangles and circles according to specifications

looking at the project you retrieved from the internet or the standards taught at the same time. It is rare for a project to not have a strong English connection. This connection is why we devote an entire chapter to literacy and math in Chapter 4. Science is also a subject that is highly complementary to math. It can be helpful to look at the CCSS for ELA and the Next Generation Science Standards for science or engineering as you are imagining your project. There are always ways to interact with literature and to write during a project so take time to think about what books your class could be reading and what ways your students will express themselves in writing. The science standards encourage students to create models and modeling is something we want our students doing in math as well.

As a final note about selecting standards, we want to make sure you understand that you are selecting standards you plan to assess. Although students may end up doing lots of standards, you do not need to assess other standards that end up being experienced in the project. This is important to understand. Projects are about students gaining a deep conceptual and procedural understanding of targeted standards. During projects, you often have students using past knowledge or even knowledge that supports the targeted standard. For example, your goal in a second-grade project could be for students to fluently add and subtract within 1000. This involves knowing how to read and write numbers to a 1000. Although this is a part of the process, you are not necessarily teaching and assessing reading and writing numbers. It is a skill that is being reinforced rather than introduced in the project.

Step 1 Applied to Complete Second Grade PBL Unit Example

In one of my classes, I (Chris) did a project about pirates. Pirates were of interest to the students I taught. I know many five- to seven-year-old children are interested in them as well. As a result, this project idea could easily be done with K–2. Second grade was chosen because they are required to understand larger numbers and burying treasure goes along nicely with large numbers. As I looked at the second grade math standards, I saw the number and measurement standards could work for my original project idea (see Figure 3.1). By the time this project occurs, students will have worked on addition and subtraction all year and this will take them to working in the hundreds position. Adding and subtracting with and without context is something that occurred over and over in my original project. The other standards deal with measurement which also occurred a lot in my project. Students measure with a ruler and compile their measurements as a class to analyze. I chose measurement to be the driver of this project because it is new math knowledge and it is the core of this project. It occurs over and over again (see Figure 3.1).

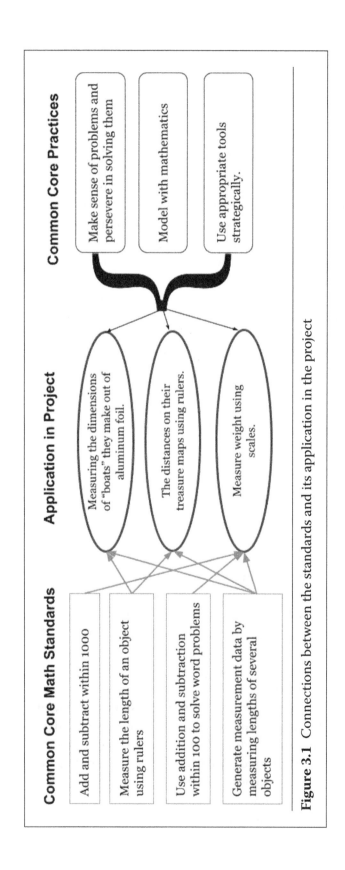

Figure 3.1 Connections between the standards and its application in the project

In this project students are also introduced to higher level concepts such as volume, density and buoyancy (which is made up of weight and volume). We feel it is fine to work with math concepts that are above a certain grade level, as long as the mathematics involved is made accessible to the students and you, as the teacher, are comfortable doing that. In this project, the teacher can just compare the two numbers (weight and volume) or the teacher can divide the two values, using technology.

Step 2: Determine the Timespan (Including When It Will Occur and Duration)

This step will vary depending on what occurred in the previous step. If you got the project from an internet search, then you had to identify the standards that align with your curriculum. After you determine the standards, then you have to find when those standards are taught. If you work from standards, then you are already looking at when the standards are taught. Once you find when they are taught, determining the timespan is easy.

To determine the timespan, you would look at your curriculum guide for when and where these standards are taught. Many districts have curriculum guides that are very specific about when standards are taught and for how long. If this is the case, you would need to decide if the time length allotted in the curriculum guide fits with what you are thinking of doing with your class. Getting too far from the district curriculum guide will cause issues later in the school year if planning isn't strategic. Timing is the key to every successful, or unsuccessful, project. Questions teachers should always consider in their planning include:

- ❏ How long should this project take to complete?
  - ■ You need to make sure you take into consideration there is some time within a project dedicated to students working on products and interacting with experts. This is done every day to deepen their understanding however it is bigger than completing a worksheet so there will be a couple of extra days.
- ❏ When and for how long are these standards taught in my curriculum?
  - ■ Many districts or states provide guidance on the order for standards to be taught. Some even provide how long to take for a set of standards. Use this to help determine length. If you don't have guidance, you can use your own personal way of teaching in the past to help guide you.
- ❏ When should I do this project so that it is completed before the state or district assessment occurs?

■ Some states have assessments that occur at different times in the year. Many districts have their own assessments to help teachers see if students are on track for the state assessments. It is important to make sure your unit will end a few days before any major assessment occurs.

❑ When would common assessments or traditional tests occur that would affect my timeframe?

■ Traditional tests and quizzes are okay in projects. In addition, many schools have common assessments as a part of their practice. This is when teachers create an assessment that all students take and examine the data. You can still give traditional tests or quizzes at strategic points to make sure kids understand basic knowledge. This must be planned for so that you don't get derailed with your timeframe.

Step 2 Applied to Complete Second Grade PBL Unit Example

The standards for adding and subtracting grows from within a 100 at the beginning of the year to 1000 by the close of the school year. Measuring using inch cubes or rulers for length and scales for weight will be new and measuring length is often found late in the school year. This project fits nicely into a two-week period and allows for a Monday presentation day which is commonly when parents come to school for events.

Step 3: Think About What Students Will Create That Will Show the Standards as Well as Evidence of Learning

When we look at our standards, we make sure to understand the verb identified in the wording of the standard. For example, the standard CCSS 2.NBT.B.9 states "Explain why addition and subtraction strategies work, using place value and the properties of operations." The verb explain is clarified with the word why and using. When you teach this standard, then you would make sure students are explaining their work as it relates to adding and subtracting. The same is true when we are planning a PBL unit that shows evidence of learning. What do the verbs say in the standard and how will the students do this during the unit? And so, we need to go even farther or deeper with our strategizing. In addition to thinking about what the students will be doing, it is important to think about who in the real world does this verb. How will we ensure that the students are mimicking the real-world work in our project? If you choose standards as your starting point, this is the time where you can figure out the idea that makes your standards come alive.

The following is an example of where a colleague, Alexandra Sánchez – third grade teacher – used this thinking in her project planning. Her project idea came from Michigan history standards and required students to understand the connection between three or more related historical events. Thinking through who must understand historical connections in real life, she decided upon a project where the class would create a museum where the exhibit showed related events in Michigan history. The students found artifacts to represent specific timeframes in Michigan history and made descriptions to accompany the artifacts. The students served as museum docents escorting and explaining the displays to the guests. The "who uses this?" part of her project ideation helped her decide the ultimate framework for her project and was well received by the school community.

At the K–2 level, it is often very easy to see how people use the math you are teaching students. As many have shared with us, the challenge is in the conceptual development and authentic learning experience. Conceptual understanding can be developed as you make connections with higher level mathematics or read books such as Chapin and Johnson's *Math Matters: Understanding the Math You Teach*. (Chapin and Johnson) We have also found doing projects developed our conceptual understanding. It is also a challenge to know how to teach math using the authentic context in conjunction with the methods you are required to teach such as ten frames. Although it is not a complete solution, check out Table 3.4 for more guidance on how to begin to rectify these challenges.

Once you think of what it looks like to achieve the standard, you select a product. Let's talk about common products and how they work in PBL. In K–2, typically all the products are made as a group. At the second-grade level, there can be an individual and a group product depending on the development of the students in the class. Typically, this would be done in the second part of the year. Regardless, there are no more than two products that students are revising throughout the project. If there are two products, one product feeds into another product. For example, in the project simulated in Chapter 2, the students made a solicitation letter which fed into the shopping day.

With PBL unit planning, there needs to be a tangible product, created by the students, that can be looked at and critiqued by an audience. Even though all things you see in the world can be connected to math, not all of them are appropriate for demonstrating mathematical learning at your students' level. We will explain options we have used at the upper levels and will explain what they look like at the lower levels. The following are options we have collected for products that can be used with most math concepts and that are routinely used in the real world:

TABLE 3.4
Building understanding of math applications

Tips for building understanding of mathematical applications in the real world
1. **Study the big ideas of math.** Sometimes we break math concepts into too many small parts and forget they connect to a big idea. Textbooks and the misreading of standards is what makes this occur. When you look at your textbook, think about what the entire chapter is saying rather than each lesson.
2. **Ask professionals how they use math.** The authors were not always math teachers. This allows us a natural ability to think about applications of math in other contexts because we did it for years before coming into teaching. You don't have to take a second job to get experience like we did. You can just ask people that you know who are not teachers how they use it. Nontraditional math careers might think they don't use math, initially. So don't stop the conversation if they say they don't use it. They may need a moment to realize that they don't use math like they were taught math in school, so they think it is not really math. Ask them questions like these: When do things that involve numbers happen in your job? Why do they occur in your job? How do you handle things involving numbers? What are the tools that you use that involve numbers?
3. **Take a course.** There are a lot of courses involving the application of mathematics. An easy starting point is to take any type of science course. Math occurs without science all of the time. However, science never occurs without math. But don't stop with science.There are a ton of courses now available for free that use math in the context of real world applications.

- ❏ A reduced version of a Viva – as Jo Boaler explains in *Mathematical Mindsets*, "A viva is the culminating exam for PhD students when they 'defend' the dissertations they have produced over a number of years in front of their committee of professors" (Boaler, 2016). We like this as an appropriate product because it includes the characteristics of mathematicians as a profession. And it can be scaled down (maturity-wise) to any level. Asking a first grader to stand up and defend the mathematics used in their project will help cement the math concepts in that student.
- ❏ Proposals – a proposal is a basic name for the written document of a possible solution. We like these because they can contain any information that relates to the solution. This is a great way to add writing skills into the project and we both believe that writing should be in every project, K–12.
- ❏ Design plans – a lot of real-world applications of math contain some element of the design process. We like design plans because when students have to justify their plans, they can bring in mathematical

concepts that demonstrate their learning. When you go through the design process students propose a solution to a problem and get feedback and then revise their solution. Then they make a final presentation. As with the "reduced Viva," the requirement of verbally defending an idea helps a student go deeper with a concept.

Step 3 Applied to Complete Second Grade PBL Unit Example

In my project, I (Chris) had students create pirate maps and calculations of buoyancy. In this project, I felt the students would be able to have two products instead of one since it is completed almost at the end of the school year. However, we thought they still may not be mature enough to do an individual product.

We decided the group product would be a presentation to family and possibly community guests. Presentations are common near the end of the school year since students have grown in their ability to speak to people. You always want an audience outside of the classroom. Gathering family and possibly community members as an audience is usually not too hard to do. The presentation would contain the following components: (1) the name of their team, the name of their boat and why that boat was selected; (2) the dimensions of their boat; (3) the volume of their boat; (4) the predicted amount of treasure their boat will carry; and (5) a short story about an adventure they had on their boat. The presentation is created as a team with each member contributing to the components.

Step 4: Build Inquiry That Ties to Each Product (Including Formative Assessments and Reflection)

There are important parts to every project. Many might not agree on any specific demarcation in the parts, but we will call them (a) the entry event also known as project launch, (b) investigating to build empathy and mathematical understanding, (c) applying knowledge to the creation of the final product(s) and (d) presenting the idea/solution/product. The entry event is so big we have a complete step for it. You will read more about it in step 6. During and throughout each of these parts students should be asking questions and learning new information or ideas or applying the learning. When this happens throughout the entirety of the project it is referred to as sustained inquiry in PBL literature.

To build the inquiry with the above parts in mind, you pull out a calendar and start with the final two days of the project which is part d (presenting the idea/solution/product). The final day is a whole class reflection and the day before that is the deadline for the final product (either the group or individual). You then plan necessary deadlines which are either revision times or portions of the products. See Table 3.5 for deadlines for the completed project in Chapter 6.

After you have placed the major deadlines on the calendar, you plan for investigations to build empathy and mathematical understanding as well as applying knowledge to the creation of the final product. You start with thinking about the empathy you want to build in students. Here are some questions you may ask yourself:

- ❏ Why are they doing this project – who cares?
- ❏ What impact will the products have on the problem being presented?
- ❏ How will you help make this the students' problem and not just a problem for other people?
- ❏ How will you scaffold the thinking around the creation of the products?
- ❏ How will students know they are creating the right answer to the problem?

Answering these questions helps to provide the framework to help you select activities or protocols to use in the project. Some of the activities will be solely to build mathematical knowledge like day three in the example project in Chapter 6. But others will be a mixture of empathy and mathematical understanding. In the pirate project, the students are the people who care. They have a natural drive for winning. On day five of the project, students are learning the math for making a boat float while also being challenged with making a boat that will hold lots of treasure to transport for the class.

TABLE 3.5
Deadlines for completed project

Monday	Tuesday	Wednesday	Thursday	Friday
		Build Boat		
Sink Day		Boat Revision		Boat Presentation Prep
Presentation to Family	Reflection Day			

You can use a lot of existing practices and current curriculum materials to develop students' mathematical knowledge. For example, having daily journal entries or daily "exit tickets" that address questions that the students still have or asking them to write about "one thing they learned today," keeps the project question or essential question at the forefront of their minds. The key to success is to take the time to formally place items on your calendar that will sustain the inquiry in the classroom.

The other thing that these calendared items do is to allow for assessing the students. Each of these daily journals or exit tickets allow the teacher to do formative assessments of the learning that is happening in the classroom. In addition, it has been shown in numerous studies how important reflection is in the learning process. We need to be overt in our plan for timely reflection by students about their progress, their group's progress and any other markers of deeper learning that may be occurring. Within a PBL unit, we sustain inquiry, formatively assess our students on a regular basis and allow for opportunities to reflect.

Once you have the major items such as deadlines and assessments on the calendar, it is time to think about the structure of each day during the project. The "PBL elementary schedule" in Table 3.1 was created from talking with several teachers who use PBL in their practice. We feel it is better to have a set time in the morning where students are given the opportunity to plan their project work, and another time at the end of the day to record what work needs to be done the next class day. To ensure students are successful with this it is helpful to have a project calendar that is shared with the students and their parents and is visible somewhere prominent in the classroom. The calendar doesn't have to be super detailed so that the students still feel like they are driving the day. Deadlines are the most common items on the calendar. During the morning meeting, make sure the class or small groups are answering these questions:

- ❏ What have we done so far?
- ❏ What do we need to get today?
- ❏ What key deadlines are coming up that we need to be thinking about?

During the afternoon meeting, make sure the class or small groups are answering these questions:

- ❏ What did we get done today?
- ❏ What do we need to do tomorrow?
- ❏ What key deadlines are coming up that we need to be thinking about?

Once the students have done their morning project planning time, teachers can make use of their content times to work on areas within the project that require a specific skill or knowledge to move the project along. If they are needing to practice addition and subtraction, for example, they would practice adding and subtracting items from the project during morning math instruction. We recommend whole class lessons occur in the morning while the project work time occurs in the afternoon.

Students do better when they are on a set schedule. However, projects can force schedule changes. A day or two prior to presentation day will need to be a project presentation planning day. It is helpful to have a special schedule for that day that you use for every project, unless you can use the existing schedule to make the planning happen. Finally, the day where products are presented might require a different schedule and it should be something planned well in advance and communicated to family and participants.

Step 4 Applied to Complete Second Grade PBL Unit Example

Authentic audiences are always helpful in keeping students focused on the fact that there is a specific date that they have to be ready to present their work. In this example, the students knew there would be parents coming to see their presentations. It is important to remember that whether you are having an authentic audience at the end or not, it is imperative to have set deadlines for students to meet throughout the project. Through the entry event students knew that they would have to build pirate boats to carry treasure to a secret destination. And when they knew that they would be showing their parents the finished boats and that they would get to sink their boats, engagement was set.

Questions about whether pirates were real and if there were any pirates around today are expected but there are some key questions students have to ask and they will rarely let you down. In this case they might ask about what they would make the boats out of and how much treasure they would get and would it all fit on their boats without sinking? You may need to keep coming back to the fact that they need to build a great boat and they need to carry as much treasure in their boat as possible. This keeps the inquiry at a high level throughout. Adding some journaling and story writing around the topic also helps with the inquiry level.

Formatively assessing each day around measuring and adding and subtracting will ensure that you have a complete picture of where students are in their learning. For example, when you hear one member of a group say they have 37 pieces of treasure and another member of their group has 43 pieces of treasure you could ask "how many more pieces do you need to also have 43 pieces?" Or when they say that they just measured the length of their boat and it was 6 inches long, you might ask them to show you how they got that.

Finally, while reading through the journals you will be able to see the level of understanding for each student.

A key to implementing this project (or any project) is to stay on track. The students work on the presentation of their boats the entire time (check out the Complete PBL Unit Example for the day-to-day look at how the inquiry unfolded). The students have small deadlines before the presentation to help keep the project on track and sustain the inquiry. They start off small with a quick creation of a boat that is a rectangular prism. This is used to help develop their understanding of boat design and how it floats. They move to a more sophisticated design informed by readings and testing. Midway through the project, they begin to find another location for the treasure and create a story explaining the buried treasure. Again, this is informed by readings. It is not a series of activities but an orchestrated program that moves students to the final product by them driving their learning. The activities serve the larger goal rather than the reverse.

Step 5: Create the Problem or Question That Will Tie the Project Together

Once you have the standards, product and inquiry, it is easier to write the problem or question that ties the project together. The problem or question is a tool used to help drive the students through the project. It is why PBLWorks calls it a driving question or challenging problem. (PBLWorks) In over four decades of teaching students math, we have rarely seen students get excited about doing a worksheet of problems. If all you do is present worksheets to your students they will, undoubtedly, be unhappy or, worse, shut down in the face of an old nemesis. You don't have to make life all unicorns and rainbows either. A happy medium is to present your students with a challenge that they know they will be able to work on over more than one class period and that they will be able to work on with others – they won't be facing this alone.

When you are creating the question for your task or project you shouldn't worry about having to produce something that will win you awards. Getting bogged down in the project idea/problem/question is one of the things that keeps people from trying PBL. Look at the sustained inquiry that you have created for your students and use it to create the question or concrete details of your project. Write a few questions and share them with others to refine it. Another method we use is taking an existing word problem that represents the skills you want to include in your problem and use it as a basis for your task scenario or PBL unit. This is not only how we come up with a problem or question, it is also how we come up with a project in general. Consider the following example from *K5Learning.com: Grade 3 Mixed Word Problems* ("Mixed Word Problems"):

Dave is working to improve the yard at his house.

1. There are two kinds of soil to choose from. If brand A is $50 for 5 kg and brand B is $48 per 4 kg, which brand is cheaper?
2. There are two pine trees at the front entrance. The tall one is 16 feet and the short one is 7 feet shorter than the tall one. Dave has a ladder that is 14 feet tall. Compared to the shorter pine tree, how much taller is his ladder?
3. Each pack of pumpkin seeds costs $8 and each pack of tomato seeds costs $5. What is the total cost for 3 packs of tomato seeds and 4 packs of pumpkin seeds?
4. He planted 5 flower beds with roses and tulips. If there are 5 roses and 9 tulips in each flower bed, how many more tulips are there?
5. Dave spent 2 hours working in the yard everyday. If he worked for 24 hours in total, how many days did he work in the yard?
6. Write the number sentence that fits this: "Dave spent $24 to buy new tools, $13 to buy seeds and $30 for a new garden hose. He spent a total of $67 at the store."

In this word problem students must compare values, add values and write number sentences. These require different skills and really don't give us anything to work with for creating a scenario that students can explore directly. However, we can use the idea of the problem as a place to create the scenario for our project. Since there are cost comparisons about soil and seeds and there is a discussion about sizes of trees, this scenario could be brought in around a school's garden or some area in the community. There might be school clubs or organizations that work with these areas and could become partners in the project. Suddenly you do have a project that could be done with your students that is authentic and has a public product.

We contend that once you have a word problem that has some depth to it, you will be able to create several possible scenarios to use as the foundation for tasks or PBL units. If you still question your imagination, then find some friends to help you brainstorm the possibilities.

Step 5 Applied to Complete Second Grade PBL Unit Example

We decided to do this project because we had seen a successful project being done by other students around pirate ships and treasure. We know there are pirate books that can be used as a catalyst and students love *Dress Like a Pirate Day*. By tying this project to *Dress Like a Pirate Day* it makes sense that we can then ask the audience to also dress up increasing the excitement of presentation day. Now we just needed to add the idea of building their pirate ship and carrying treasure. To combine the understanding that ship loaders have to worry about carrying too much cargo with the idea that pirates are always taking treasure and moving it around we created the following project question: "How do we, as pirates, carry our treasure safely?" And for this book we offer an alternative where we are actually addressing the shipping industry: "How do ship loaders get their goods to places safely?"

Step 6: Decide How to Kick It Off

Once you have a problem for students to work on you will need a way to get students interested in solving it. In many inquiry methods, this is known as the "entry event" or project launch. The key is to heighten the interest of your students, so feel free to do whatever it takes. In addition to heightening interest the entry event should provoke questions leading to a teacher-facilitated discussion of what questions the class will want to explore during the project or task days. For kindergarten or even first grade students, the entry event takes a few days. This is because their memory of an experience is so short lived. Plan for the first three days to be focused on the students getting engaged in the project.

What makes a great entry event? We have seen real people, with real problems, come in and talk with students or virtually with Skype/Zoom. We have seen people impersonating real (and fictional) characters who ask the students for their help. We have seen videos, letters, and emails used as requests for help. Field trips can be great sparks of inquiry as well as anything anchored around the class pet. The key to deeper learning is increased authenticity. This means that real people with real issues will always be the best way to reach higher student interest. However, that is not always practical. And authenticity may require a longer time frame and a higher level of mathematics than you have at your disposal. These are things that will have to be weighed in the planning of the calendar.

As you've seen, we keep coming back to the calendar or time frame that will be used for the task or project. Timing is so important and will lead to the success (or non-success) of the task or project. Therefore, in your first few tasks or projects, we recommend keeping entry events to a letter or email explaining what needs to be done. The advantage to being in an elementary setting is that there are, often, parent groups ready and willing to help teachers in whatever project you come up with. Start simple so that you and your students feel successful at the end of the task or project. Then you and your students will be open to doing more of these PBL units. You will find yourself wanting to spread your wings to bring in a community member with a problem that they need help with. Your projects will instantly be more authentic and your students will find themselves going deeper in their learning.

Step 6 Applied to Complete Second Grade PBL Unit Example

For this project, we would have students reading a book about pirates. One book we might read is *Shipwreck on The Pirate Islands* or another pirate book where a pirate's ship sinks or wrecks in the ocean. We would read the book with great inflection and the teacher might even dress like a pirate. This is followed by students looking at pirate maps and then treasure being found by teams of students. Pirates and buried treasure is a common scenario children play out together. Using kids' imagination in general is a great tool to engage them. Their treasure would be pennies that they have to count which adds to over 100 for each team.

The hardest step in the preparation of the entry event is having enough treasure so that each person has a large number and that each group would have at least 100 pieces of treasure. We recommend pennies but we know that teachers love using places like Oriental Trading Company or Party City to get inexpensive trinkets. Should you choose pennies, you will only need around 1000 pennies (10 dollars worth). And, since they must count you can just break the pennies into 6 fairly even piles (one for each team) and then have each group break that amount into 4 fairly equal piles (for each member of the team). Since the amounts for each team and each student is not known, you can just tell them that the total will add up to 1000 (or the amount you end up having for the project).

References

Ainsworth, Larry. *Power Standards Identifying Standards that Matter the Most*. Houghton Mifflin, 2003.

Chapin, Suzanne, and Art Johnson. *Math Matters: Understanding the Math You Teach*. Second ed., Math Solutions Publications, 2006.

"Mixed Word Problems." *K5 Learning.com*, n.d., https://www.k5learning.com /free-math-worksheets/third-grade-3/word-problems-mixed/4-operations -mixed.

PBL Works. "Gold Standard: Essential Project Design Elements." *What is PBL Gold Standard Project Design*, n.d., https://www.pblworks.org/what-is-pbl /gold-standard-project-design.

CHAPTER 4

Math and Literacy

Two Sides of The Same Coin

Before we review how math and literacy are more alike than different, let me expand on my story mentioned in the forward. I, Telannia, cannot remember which love I discovered first: a love of reading or a love of logic puzzles. Yes, logic puzzles. The ones near the checkout stand at pharmacy stores or in the magazine aisle at the bottom because no one hardly gets them. I would literally escape in them. I excelled in school including language arts and math. It wasn't until high school that I ran into my first challenge with the subjects. It was when I attempted AP classes. Although I was editor of the newspaper and a part of the yearbook, my AP English teacher told me I didn't have the writing skills for AP and bumped me back to junior level English and never let me return. In my calculus class, my papers would almost look unrecognizable at times. I begged to get out but the teacher refused and told me I was going to be fine. In the end, I scored a 4 out of 5 on the AP Calculus test. Despite the setback in language arts and the great accomplishment in math, I pursued my dreams of a career in journalism. I had a wonderful career in print media using both language arts and math skills.

When I came into teaching, I thought about my K–12 experience. As an African American female, I had a roller coaster of an experience. I wanted to tilt the scales for other students of color so that hopefully I could help increase the great experiences they could have in school. I thought about how good I was at most subjects. I realized I could teach any of them as long as I took the tests. So, I thought to myself, if I could teach anything and were to make the most impact for students of color, what should I teach? What subject has the most need of support and is often a bad versus good experience?

As a result, I chose to teach math. It is a subject that too many people feel they are not capable of doing. It is where people have their worst experiences. When people would share, so proudly, how they were not math people, I would say that I have a degree in journalism and made a high score on one of the most difficult math exams. I was proof that they could do both. They only needed the proper support. For over 15 years, I have worked to unpack the reasons why I never had a problem with language arts or math. Why I never saw myself as a person who was only capable of doing one thing well. I spent years thinking about my thinking and natural ways of being as well as studying the best practices of teaching and learning. My biggest revelation was that I don't think of them as two different things but as two sides of the same coin. This chapter is not only my brain opened up to help you make the connections I made, it is also the result of books, professional development and people who have helped me better understand what my brain was always doing growing up.

In order to explain the connections, this chapter is divided into two main sections: macro connections and micro connections. In macro connections, I make broad connections between language arts and math. In micro connections, I give more narrow details. Another way to think about it is that macro is the forest while micro is the trees. I use literacy knowledge such as phonics and comprehension strategies to connect to mathematical strategies. Neither the macro nor micro include all the connections that could be made. These are just some of the major connections I have found over the years.

Some people like to see the big picture (macro) before they see the small details (micro). Some people are the opposite. Feel free to read this chapter in the order of your preference but definitely read both. It takes both views to understand the connections. Let's get started!

Macro Connections

There are four macro connections explained in this section. All of them are big ideas of language arts and mathematics. After each idea, there is a short explanation of how this connects to your practice. The ideas are based upon the principles of each subject area as well as the human condition. It consists of ideas everyone is familiar with. It shows how mathematics and language arts are indeed two sides of the same coin.

Reading, Writing, Speaking, and Listening Are the Parts

Language was created to enable us to be in relationship with one another. It is how human beings communicate what we see, think and feel. At first, we communicated only verbally. The Smithsonian Museum of Natural History says that written language was formed around 8,000 years ago, with true forms of writing taking the next few thousand years to develop. (Smithsonian National Museum of Natural History). Language is broken into four major parts: reading, writing, speaking and listening. Even though Common Core State Standards raised speaking and listening to an equal position as reading and writing, most classrooms still spend a large amount of time teaching students how to read and write. It is the first love of most teachers, especially K–2. To understand the connections between the two subjects, let's explain the four parts in a more general form:

- ❏ Reading: the ability to understand ideas and stories communicated in a written form.
- ❏ Writing: the ability to communicate an idea or story in a written form.
- ❏ Speaking: the ability to communicate an idea or story verbally.
- ❏ Listening: the ability to understand communication that was in oral form.

Mathematics has the same four parts. In math, students still read to understand ideas and the underlying story. The ideas and stories are focused on the quantitative side of our world. Students must learn how to read numbers and the quantitative description of the world in which numbers reside. Writing is the same. The focus is on how to communicate those quantitative stories. Just like reading and writing go hand in hand, speaking and listening are matching pairs as well. It has proper structures of the way we say quantitative relationships verbally and thereby allow us to hear the idea or story clearly.

You see, mathematics completes the communication process. There are occurrences in the world that just words cannot communicate clearly. For example, how do I communicate how much I have of something or how much I need of something? While language's foundation is a way to be in relation to each other, mathematics is a way to be in relation with the world around us. It is a way for us to make sense of the world as well as an additional way to communicate it. I like to call it a universal language. Just like traditional language, the ability to read and write mathematically is the most common focus. Students spend most of the time learning how to read and write numbers and symbols rather than equally learning how to speak and listen to it. For both language and math, all four are key to mastery.

Application to Practice

As you approach the teaching of mathematics, remember you are still teaching reading, writing, speaking and listening. You are helping students grow in understanding the world around them through a quantitative lens. You are helping them to communicate just like you are with letters that form words. As a result, it is helpful if you use language that shows how they are two sides of the same coin. You always use the words reading, writing, speaking and listening when you go through language arts lessons. Do the same when you are doing mathematics. For example, when you review numbers, share how it is similar and different from letters. There are 26 letters and there are ten numbers. The letter "O" and the number 0 look similar but they are not the same. Just like letters can be alone or put together to mean different things, the same happens with numbers. Share how they are always learning how to read. Sometimes they are reading letters and sometimes they are reading numbers.

Symbols Are Their Backbone

Backbone is a common word for foundation. It is figurative of how everything is connected to this one piece. Both language arts and mathematics use symbols as their foundation for learning. As alluded to in the first macro connection, reading and writing uses symbols to communicate. Like most languages, language arts uses letters as the primary symbol for communication. Mathematics uses numbers as their primary symbol. Everyone uses a great deal of their formative years learning both symbols. You can probably remember being a child or singing to your child different songs that help you remember letters and numbers. All 26 letters and the first ten numbers are often known by two- or three-year-olds. It typically just depends on how quickly a child starts to communicate complete sentences with other adults.

Basic letters and numbers are not the only common symbols that are the backbones of the subject areas. They have advanced symbols that help support the basics. For example, letters also include symbols for punctuation like periods and commas. With the advancement of technology, emojis are starting to take on another form of symbol support. It started with text messaging and has spilled over into formal forms of communication like email. In contrast, mathematics has symbols to support its communication. The most common is operations like addition, subtraction, multiplication and division. Just like punctuation provides clarity to words, operations provide clarity to numbers. There have also been extensions made to the symbols of mathematics as well.

Computer science is based upon mathematical principles and has therefore extended its symbols.

Application to Practice

Think of letters and numbers together rather than separately. Think of letters as the way to communicate everything that is not quantitative in nature while numbers allow us to complete our communication needs when representation of a value is needed. Explain to students that language is a collection of symbols rather than just letters. Share how they are learning a language which requires them to learn its symbols and how they interact. Staci Walling, a kindergarten teacher in Oklahoma, shared how she writes number sentences with words under the symbols on the board. For example, she would write 2 + 3 = 5 on the board but the word "and" would be under the plus symbol while the word "is" would be under the equal symbol. Doing this is exactly like having pictures connected to words. It is a necessary and powerful way for students to become fluent in math.

Rigid and Flexible Structures

It was not until college that I discovered that some rules could be bent or even broken when it came to communicating. It was my junior year and I was starting to take courses in my degree area. One of my professors said we have to learn the rules of newspaper writing. I was excited because I had already been learning the rules since middle school. However, he floored me when he said once you learn the rules then you understand how they can be bent and elevate your work. I never knew you could break the rules in language arts. It was several years later that I realized you can do the same in mathematics. For example, order of operations is not a rigid process that you must follow. There are other orders that you can do and it still works. Initially, kindergarten students believe you must start at 1 to count. Counting on from another number or skip counting is like breaking the rules to them.

As we learn language arts, we are taught all the structures to follow. Early in our development, we learn a sentence has a noun and verb. As we continue, we learn more structures to a sentence. Similarly, in mathematics we learn specific ways to solve a problem. In both instances, the way we are taught gives the impression that these ways of doing things have to occur exactly as we were taught. For example, in math, there is only one way to solve a problem. It is another common rule that is misunderstood when only one way is shown over and over again. Both subjects have structures that are both rigid and flexible. It is vital that students learn both to properly grow in proficiency in both.

Application to Practice

We must be very careful when we teach the structures of language arts and mathematics. No student should discover the flexibility within their learning when they leave the K–12 setting. It should be constantly a part of their understanding. This most commonly happens when students have limited or no experiences with inquiry in their learning. A simple way to accomplish this is by always using students' prior knowledge to connect to their new knowledge. For example, in language arts, it is common to use prefixes of words to help students learn new words. In mathematics, you can use students' knowledge of the order of numbers to learn numbers after ten. An advanced way to implement this practice is to actually have students do an inquiry-based activity. You can see Chapter 5 for some examples of inquiry activities.

Concepts Grow Over Time

As mentioned in the first macro connection, language arts and math are both broken into reading, writing, speaking and listening. Even before the standards movement, these breakdowns for language arts were its foundation. When standards were introduced, the same structure was used to communicate what students should know and do within each section. As you look at the language arts standards it is really easy to see how the concepts grow over time. Table 4.1 shows how one reading standard of finding key ideas and details for literature grows from kindergarten to second grade.

Mathematics grows over time as well but it is harder to see especially in the way the United States organizes and teaches mathematics. I (Telannia) believe there are two reasons why people have a hard time seeing the concepts grow. Reason one is we don't develop students' understanding conceptually. This is changing but still too many experience mathematics as an execution of procedures rather than the study of numeric-based relationships. The second reason is the way we divide mathematics into different names at the higher level. For example, in other countries, after students complete middle school, they still

TABLE 4.1
Key ideas and details

Kindergarten	First grade	Second grade
RL.K.3 With prompting and support, identify characters, settings and major events in a story.	RL.1.3 Describe characters, settings and major events in a story, using key details	RL.2.3 Describe how characters in a story respond to major events and challenges

TABLE 4.2

Language arts compared to mathematics

Language arts			Mathematics		
Reading			Reading		
Literature	Informational text	Foundational skills	Numbers/ quantity	Representations (tables, graphs, sentences, etc)	Shapes
Writing			Writing		
Opinion	Informative/ explanatory	Narrative	Numbers/ quantity	Modeling (tables, graphs, sentences, etc)	Shapes/ measure- ment
Speaking and listening			Speaking and listening		
Comprehension	Collaboration	Presentation	Comprehension	Collaboration	Presentation

take mathematics. They don't take algebra I, geometry or algebra II. It is mathematics and then a grade level number after such as mathematics 9. Sometimes there is a description before mathematics such as discrete mathematics. This distinction we have in the US causes people to further feel like they are a type of person to the point where they pick a particular type of math. You may even recall feeling like you were good at algebra but not geometry or vice versa. It is all mathematics, just different lenses.

To help see through this lens for mathematics, let's look at breaking down mathematics through the four parts that language arts is broken into. When we see how language arts are broken down, it is easier to see how one concept is the same, such as reading, but it may have parts that grow in sophistication. Table 4.2 shows the breakdown of language arts with math next to it in the same breakdown using the Common Core State Standards.

Remember how Table 4.1 showed the development of students' ability to identify key details with more and more sophistication. With the connections shown in Table 4.2 in our mind, let's look at how a particular portion within mathematics grows over time. Table 4.3 shows how counting is one of the ways we "read" in mathematics. Students first understand the composition of a two-digit number. They develop an understanding of base ten. In first grade, they use their understanding to count up to 120, now seeing how two digits become three digits. In second grade, they solidify their understanding of three digits numbers by assigning place value.

Application to Practice

There is a high urgency in most elementary schools to make sure students get a strong foundation in language arts because it helps them in later grades when

TABLE 4.3
Mathematics number sense over time

Kindergarten	First grade	Second grade
K.NBT.A.1 Compose and decompose numbers from 11 to 19 into ten ones and some further ones, e.g., by using objects or drawings, and record each composition or decomposition by a drawing or equation (such as 18 = 10 + 8); understand that these numbers are composed of ten ones and one, two, three, four, five, six, seven, eight or nine ones	1.NBT.A.1 Count to 120, starting at any number less than 120. In this range, read and write numerals and represent a number of objects with a written numeral	2.NBT.A.1 Understand that the three digits of a three-digit number represent amounts of hundreds, tens and ones; e.g., 706 equals 7 hundreds, 0 tens and 6 ones. Understand the following as special cases: ❑ 1.a 100 can be thought of as a bundle of ten tens – called a "hundred" ❑ 1.b The numbers 100, 200, 300, 400, 500, 600, 700, 800 and 900 refer to one, two, three, four, five, six, seven, eight or nine hundreds (and 0 tens and 0 ones)

the text gets more complicated. A strong foundation is needed in mathematics as well. Remember they are two sides of the same coin. Being proficient in both is what makes a literate person, not just language arts. Even if you do not understand how it connects to higher level mathematics, knowing that it is a piece along a continuum of learning is a start. As you work to prepare a lesson, look at what happened a grade level below you and what is going to happen two grade levels after you. This will help you see the bigger concept and how it grows. Looking at various grade levels is a part of the vertical alignment process. This is often done at the district level with invitations open to teachers to participate. Teaching Channel has a resource to help you understand the basics of aligning your curriculum. (The Teaching Channel)

Micro Connections

There are three micro connections explained in this section. Again, these are the specific details within language arts and math that connect. These connections are sometimes about the details of the structures and other times it is about the strategies. There are many micro connections. This section mentions a few that I believe will help you the most. I chose them because they represent different levels necessary to become a successful reader. Unlike the macro section, there is no short explanation of how this connects to your practice. This is because it is directly in the explanation of the connection. Like the macro connections section, this shows how mathematics and language arts are indeed two sides of the same coin.

Symbols and Their Combinations

Symbols are the bedrock of being able to read. The thought of 26 letters being interchanged to make words is an amazing idea if you think about it. The same amazement occurs with numbers, however the combinations are using the symbols 0 to 9. It is with this basic structure that both language arts and mathematics begins. Remember how every parent starts their child off with learning the letters of the alphabet and counting from 1 to 10. Children come to school and we reinforce this baseline understanding and grow it by showing them the wonders of how these symbols combine.

In language arts, one of the strategies to help students understand letter combinations is called phonics. The National Literacy Trust explains how phonics is a way of teaching how to read and write. They say, "written language can be compared to a code, so knowing the sounds of individual letters and how those letters sound when they're combined will help children decode words as they read" ("What is Phonics?"). As literacy expert Nell Duke states, the need to teach letter-sound relationships is settled science (Duke). Phonics is sometimes an actual section of the day rather than just incorporated during the reading block in a school day. I like to think of it as the skill you need to play the game.

In mathematics, the goal is the same as phonics – the ability to read and write. It is just that they are working to be able to read number connections. You are working to develop their understanding of place value which is how the numbers combine rather than letters. Place value is how numbers come together to form different words. For example, students understand first that just the numbers 0 to 9 represent the "one's" place. However, numbers written next to each other means you have to look at the place value of each digit to determine how to say it.

Similar to phonics the practice is primarily the focus of grades kindergarten to second and it is reflected in the standards. If you look at Common Core Standards, you would find it connected within counting and cardinality and Number and Operations. This is because to understand place value, you have to understand how numbers can be composed. For example, in kindergarten, the only common core standard for number and operations in base ten is for students to understand how to compose and decompose numbers from 11 to 19. The emphasis is on breaking them into 10s and 1s. This allows students to see the basic symbols are 0 to 9. When you have an amount over 9, then you represent it by using a 1 in front of the additional amount you have. The 1 then means you have 1 set of ten and "some further ones" as the standard states.

Sentences

Sentences are words put together with punctuation. Mathematics has sentences as well. It is when numbers and symbols are put together. For years, I have studied the way mathematics is taught at the elementary level. I almost did cartwheels when I first saw the word "sentence" used in an elementary math textbook. It basically was showing teachers that they should ask the students what the sentence is for a particular problem. At that time, I had been directly telling my students that an equation or inequality is a sentence just like sentences in their language arts class. I had wished that the language I was using would be used at earlier grades so that it was not weird when they arrived at the high school level. In language arts, letters come together to make words which combine with other symbols and they make sentences. The same occurs in mathematics. Let's look at an example of how to make the connection of English sentences to math sentences.

Kindergarten students are beginning to work on how adding is putting things together while subtraction is taking things apart. The Common Core Standards divide this concept into five standards. The first standard requires kindergarteners to represent addition and subtraction with a lot of representations including equations. You start by activating the students' understanding of a simple sentence that is also connected to a picture like most children's books are organized. An example would be a picture of a cow jumping over the moon with the sentence "The cow jumped over the moon" written underneath the picture. You ask the students to read the sentence with you. Explain how the sentence is exactly what is happening in the picture. Point to the word cow and the picture of the cow. Show how the letters cow come together to represent that specific animal. Point to the word moon and the picture of the moon. Share how it is connected like the cow. Ask them if there is another word that connects to the picture. They should be able to connect the word jump.

Then have students look at a picture where kids are sharing candy. In the picture, one person has four pieces of candy while another has two pieces. They are pushing their pieces of candy together. Share with the students how they can have a sentence that connects with this picture as well. It will be a math sentence. Below the picture, is written 4 + 2. Explain how the 4 represents the person on the left with four pieces. Ask the students where they see the 2 in the picture. Then ask the students what they think the plus symbol means.

They should be able to see how the movement of students moving their candy together is connected to the addition symbol. There are not a lot of resources we found in helping build this type of understanding. However, it is a simple adaptation to any resource you already have such as picture books and story problems.

Comprehension

It is one thing to be able to say a word or sentence. It is another thing to understand what a group of sentences says. This is even more complex when there are no pictures to help support your understanding of a book. As a result, we spend most of our class time helping students comprehend. In a video, Dr. Nell Duke explains effective comprehension strategies such as monitoring and making inferences (*Dr. Nell Duke Explaining*). Comprehension is key to reading a book with or without pictures. Good readers employ so many strategies at once. These strategies are often taught heavily in grades K–2 and reinforced in later grades. It is vital for these strategies to be explicitly taught and exercised by students using specific and engaging text. Effective teachers of reading comprehension do the following:

- ❏ Set the right conditions for students to need the strategy.
- ❏ Use a compelling text that best models the strategy.
- ❏ Model the strategy.
- ❏ Allow students to use the strategy with support of the teacher.

The good news is that many of the strategies for reading comprehension are exactly what mathematicians do. Just like in developing students reading comprehension hinges on a compelling text, mathematics hinges on a compelling problem. Table 4.4 includes some of the strategies Duke mentions in her video about language arts versus mathematics. Think about how the examples in the diagram connect to your current teaching of language arts and mathematics. If they don't connect, how could you incorporate them into your practice?

Although we do not have research-based evidence, we have found in our years of teaching that students who are challenged with comprehending in language arts also have problems comprehending in mathematics. As a result, supporting students with strategies for comprehension starting at kindergarten and continuing through second grade is vital. It is really important to incorporate other details that help support the strategies above. For example, visualization has a lot of components within it when it comes to language arts. For mathematics, it means you have flexibility in the way you "see" numbers. Number talks is one way to develop students' ability to visualize mathematics mentally.

TABLE 4.4
Language arts compared to mathematics strategies

Strategy	Language arts	Mathematics
I wonder (self-question-ing)	Students think about what they wonder while reading a text. They are specifi-cally taught the phrase, "I wonder." For example, while reading students say I wonder why the character did that	As students review a rich math problem, they should be asking questions. For example, they should say I wonder what I am trying to figure out
Monitor	Students are paying atten-tion while they read if it makes sense. If it doesn't, they employ a strategy like rereading the text	As students review the problem, they are making sure they understand what the problem is asking. If they don't, they employ a strategy like rereading or making a visual picture
Activate prior nowledge	Students before reading should think about what they already know about the text. They should make predictions based upon this prior knowledge	As students face problems with diagrams or math symbols only, they have to think about how they can use what they know to solve the problem
Make inferences	Students should read between the lines of the text they are reading. They should make connections between different ideas in the text so that they can make a conclusion that is not directly in the text	Real math problems don't always give the stu-dent direct information. Students need to be able to arrive at conclusions based upon the information that is given
Visualize	Students should make pic-tures in their mind as they are reading. They need to think of the book creating a movie in their head	Creation of pictures is the best strategy for mathemat-ics. Students need to be able to convert a problem into a visual picture i.e., rods or number lines

References

Dr. Nell Duke Explaining. Performance by Nell Duke, Youtube, 2013, https://www.youtube.com/watch?v=CgSRH0EYvhU.

Duke, Nell. "Phonics Faux Pas." *AFT.org*, n.d., https://www.aft.org/ae/winter2018-2019/duke_mesmer.

Smithsonian National Museum of Natural History. "What Does it Mean to Be Human." *What Does it Meant to Be Human*, n.d., https://humanorigins.si.edu/human-characteristics/language-symbols.

The Teaching Channel. "Aligning Curriculum Across Grade Levels." *Aligning Curriculum Across Grade Levels*, n.d., https://learn.teachingchannel.com/aligning-curriculum-module-sac.

"What is Phonics?" *Literacy Trust.org*. National Literacy Trust, n.d., https://literacytrust.org.uk/information/what-is-literacy/what-phonics/.

CHAPTER 5

Key Teaching Strategies for PBL

In Chapter 3, we said one of the steps in planning a PBL unit is to build inquiry that connects to the product creation. Inquiry is not always a common practice in the United States when learning mathematics. However, inquiry-based strategies are prevalent in K–12 classrooms and recent research shows that these strategies improve learning. These strategies can be applied in a mathematics context. A term we are seeing related to inquiry-based strategies is thinking-based classrooms. These are classrooms where thinking routines are the norm. Students in these classrooms are often working collaboratively and the routines help groups interact with their work. Most of these routines or strategies can be applied to all content areas.

This chapter highlights some of the routines and strategies we have used in our own classroom or have seen being used in other classrooms. All of them have been used in classrooms from Pre-K to post Secondary. The strategies are designed to enhance mathematical instruction and can be used on a daily basis. We will give a brief overview of each strategy or routine and discuss how it might be used in your classroom. Some of these strategies we consider essential in PBL units and others are optional. Our favorite strategies for implementing projects are marked with an asterisk. The chapter ends with how to plan lessons in a project unit with these strategies in mind.

DOI: 10.4324/9781003237358-7

Strategy 1: Investigation*

Too often we begin teaching mathematics by showing students what they are going to do rather than letting students make sense of mathematics through a rich investigation. An investigation can be students completing guided questions or an engaging problem that can be resolved in one class period. They can be completed individually but it is more powerful when students work collectively. To be successful in implementing this type of strategy, you need a structure for students to follow.

There are many structures out there that support you doing an investigation with your students. When Common Core was growing, the Mathematics Assessment Project created a resource that has engaging problems and ways to implement it ("Welcome to the Mathematics Assessment Project"). In his book, *Building Thinking Classrooms in Mathematics*, Peter Liljedahl goes into great detail of how to use engaging problems before formal teaching of a mathematical concept (Liljedahl). This is a great resource if you want details of how to implement it in your classroom. Many textbooks have great explorations before the unit begins. They often give lots of guidance on how to implement a concept with students. There are also investigations where the structure is built around the use of a particular calculator or manipulatives. For example, Texas Instruments created a section on their website (education.ti.com) that has an investigation for every concept (Texas Instruments). There are many manipulative providers such as Hand 2 Mind ("Teacher Resources and Classroom Supplies"). All of them provide support on how to use their tool as a way to investigate mathematical concepts. This strategy is a perfect replacement or addition to teaching the content in a project.

Resources

- ❏ Materials by Peter Liljedahl (available at buildingthinkingclassrooms. com/).
- ❏ Collection of 3-ACT tasks (available at gfletchy.com/3-act-lessons/).

Strategy 2: Protocols*

A protocol is a structure to work with other people to learn content, create a product or refine a product. The structure helps prevent the common pitfalls that happen when people work together. It is essential to have a way for students to work together in a project. Basically, it is a process that often includes specific times attached to the directions. You are probably even familiar with some protocols. One you may have done is a chalk talk. Chalk talk is where students silently write in response to a prompt that is the center of a piece of paper. This is a really popular protocol to activate prior knowledge or to see what knowledge students have after instruction.

Sometimes the protocol is about developing understanding of the concept while building a soft skill like critical thinking. When a protocol develops a particular way of thinking, author Ron Ritchhart, calls it a thinking routine. In a few of his books, one of which is *The Power of Making Thinking Visible* (Ritchhart and Church), Ritchhart explains how thinking routines foster deep learning, cultivate engaged students, change the role of teacher and student, enhance formative assessment, improve learning and develop disposition of thinking.

Resources

❑ Project Zero's Thinking Routines (available at pz.harvard.edu/thinking-routines).

❑ School Reform Initiative Protocols (available at schoolreforminitiative.org/protocols).

Strategy 3: Seminars or Fishbowls

In a seminar model, the teacher leads students through a discussion of the topic being examined. The most common form of seminar is a Socratic Seminar. Socratic Seminars are designed for students to help each other understand at a deeper level. The teacher becomes a facilitator and leads the discussion by posing provocative questions. To reach the highest level of discussion students must come into the Socratic seminar having prepared themselves by answering a complex problem, reading an informational or provocative text, or they are at a crossroads about a concept. Regardless of the nature of the discussion, a seminar is used to develop a deeper understanding about an idea or concept. Seminars also give students a chance to present and critique alternative methods of solving a problem or ways of thinking about a solution. Seminars can be a great way to introduce a project or deepen the understanding while finding a solution to a project.

Another seminar model that is more intentional with the reading, writing, speaking and listening cycle of learning is the Paideia seminar ("The National Paideia Center"). As the National Paideia center states, a Paideia seminar centers around a text (tangible items such as a poem, painting or a math problem). It has a specific set of questions that is a combination of Socratic and maieutic questions (relating a topic to the students' lives). Finally, it has specific speaking and listening goals that students engage in before and after the seminar. This type of seminar is an excellent way of deepening the understanding of a concept during a project, developing students' ability to communicate effectively and ability to work collaboratively.

A fishbowl is a discussion strategy that has many purposes and modifications. The basic structure is to have a group of students discussing or completing a task together while the rest of the class is in a circle around the group. The students on the outside of the group can be taking notes, completing a checklist or just watching to answer questions later in a discussion. A common purpose of the fishbowl is to model expectations or discuss a problem or concept. This strategy is an excellent way to teach proper mathematical communication or teamwork during a PBL unit or deepening understanding of a concept.

Resources

- ❑ Example of fishbowl in action (available at edutopia.org/math-social-activity-cooperative-learning-video).
- ❑ Padiea Seminar explained (available at paideia.org/paideia-in-action/index).

Strategy 4: KWL or Problem-Solving Method*

The KWL chart is a simple and familiar tool for educators. Students identify what they already know, what they want or need to know and then finally what they learned. It is a scaffolding strategy to break down the learning or problem-solving process. Some people have adapted the KWL model. An adapted version was used in Chapter 2 – the Know and Need to Know list. Although it is a common method used by teachers, KWL is not the only method to use for a PBL unit.

Co-author Telannia Norfar created the "ZOOM" process where students think of problem solving like the lens of a camera. After students are introduced to the situation, students begin to fill out the first three sections of the zoom process form. The first section is students writing essential things they know about the situation. Students then formulate the components of the solution. Students end with initial questions about the situation. Students work back and forth between the third and first section as they find answers to their questions and write new questions. When students feel like they have a possible solution, they complete the final section and revise it as their thinking evolves.

EL Education's Two Rivers Charter has a problem-solving method called the KWI (EL Education). The K stands for what they know. The W stands for what they need to know. The I stands for ideas. This method is reinforced using a form where the problems are printed at the top and the method is at the bottom divided into three sections with a prompt for each section. The first section prompt is "What I know about this problem." The section prompt is "What I need to figure out.. And the final section is "Ideas for solving." Students complete this form every time they are solving a problem. This allows for reinforcement of the process for solving any problem.

Resources

❏ Zoom Process (available at mathpbl.com).
❏ Example of KWI in action (available at vimeo.com/117861347).

Strategy 5: Use a Lesson-Structured Model*

Many teachers see inquiry as a messy, unstructured process. However, this is farthest from the truth. Inquiry follows a very structured process so that unstructured learning can happen. The following structured lesson models can be used in and out of PBL. They are great structures to add to your tool chest of strategies.

The *workshop model* is one frequently used. It is a basic three-part structure where you have an introduction, work time and debrief. The introduction is where the teacher explains what will occur during the work time. It is just enough to get students started. It is typically no more than 10% of the class time. The introduction is followed by the work time. The work time is for the majority of the class time or 70 to 80%. The students are making sense during this time. The teacher is walking around checking in on students' progress and collecting students who will participate in the debrief. Sometimes teachers will do a whole class comment during the work time but not so much that it takes up a lot of the students' work time. The debrief is students presenting their understanding.

The *5E model* is similar to the workshop model but is more structurally defined. The E's are Engage, Explore, Explain, Elaborate and Evaluate. As explained by NASA, this is "a teaching sequence that can be used for entire programs, specific units and individual lessons" ("The 5E Instructional Model"). They explain the process in terms of a lesson. Engage is the section of the lesson where you get students interested while also assessing prior understanding. The Explore stage is students learning about a particular topic in a way where they are building their own understanding. Once students have gained some understanding, they move into Explain which is an opportunity to communicate what they have learned so far. It can be written or verbal. They can share it with other students, experts or the teacher. After Explain, students move into Extend which sometimes does not occur depending on the lesson. This is where students use their new knowledge and continue to explore its implications. The model ends with Evaluations which is often said to really occur all throughout. It is where students and teachers determine how much learning has occurred.

You can think of 5E as a way to plan a PBL unit as NASA suggests. When overlapping this model with the PBL process, the Engage portion is an activity that starts the project. It also can be used to re-engage students within the PBL. During the Explore, Explain and Elaborate parts of the model, students are working to the solution. And during the Evaluate section students can be explaining their solutions or their steps to a solution. For those of you who already use the 5E Model, you may want to think of planning a PBL with that lens in mind. It will help you make understanding the PBL planning process a fairly simple shift.

The third structural option is from the book *The 5 practices for Orchestrating Productive Mathematical Discussion* (National Council of Teachers of Mathematics). The five practices are anticipate, monitor, select, sequence and connect. In this model, you start with an investigation where students would have multiple ways of seeing the outcome. *Anticipate* is where a teacher takes the time before giving the investigation to think about all of the ways a student could do the investigation. The teacher gives the investigation to the class and *monitors* the students working. This is so they can select students who will present. This selection is very targeted so the teacher can sequence the presentations in a way that students make *connections*.

Resources

- ❏ Workshop model in action (available at youtu.be/UkloO4DAIpo).
- ❏ Orchestrating a discussion explained (available at youtu.be/Ad8gcL89 V3c).

Strategy 6: Journaling*

Reflection is a key element in PBL therefore a journal is a great practice to have as a standard in your classroom. The journal can be combined with an interactive math book or it can stand alone. The key is to have students think about their learning regularly and in a safe way. In journaling students have the freedom and the safety of writing down their ideas, explanations or solutions. At certain points during a PBL unit teachers set a time for reading the journal entries and there are, usually, additional times for peer evaluations of the journal entries. Sharing the journals can keep students accountable to the process and to each other. This can be especially effective when students are working in collaborative groups. Some digital options for a journal include Google docs, Sown-To-Grow website and blog sites.

Strategy 7: Question Formulation Technique (QFT)*

When we started our journey to incorporating inquiry in our classrooms, we found out quickly that students needed help with questioning to learn. It is not something we normally teach students to do. This particular strategy was developed by Dan Rothenstein and Luz Santana. They are co-directors of the Right Question Institute (Right Question Institute). It is another great technique in and out of projects. It provides a great way for students to create questions. Here is an overview of the process that is described in their book, *Make Just One Change*, and on their website:

- ❏ Students generate questions from an image or a sentence following four rules.
- ❏ Students define the questions as open or closed.
- ❏ Students change an open question to close and vice versa.
- ❏ Students select the best questions to guide their learning given a task.
- ❏ Students reflect on the process.

Strategy 8: Word Wall/Project Board*

Most elementary classrooms have a board that is dedicated to the learning for a unit or theme. Word walls, a place where high frequency words are displayed, are often incorporated with this wall. It is vital for your classroom to have dedicated space on a wall where project information resides. However, it is not just for visual appeal. It is a living document that you and students interact with in a project. As Ian Stevenson, a former National Faculty member for PBLWorks states,

> One "move" I made that enhanced my PBL teaching was developing the classroom project wall into an active teaching tool,

rather than a bulletin board. The project wall is the visual space in the classroom that helps manage information, project questions, calendars, standards, assessments, and resources that guide student learning during the project.

(PBLWorks)

Strategy 9: Collaboration-Based Tasks/Activities*

Collaboration is a necessary skill set for life. Collaboration is not just getting students to effectively work in a center. It is truly having students work together in an interdependent way. Being effective collaborative partners requires a different set of skills. Collaboration requires sharing of ideas, commitment to a shared goal and constant negotiation. These are skills that must be taught, modeled and assessed. You must include collaboration-based activities that help students learn and improve their work. Let's look at how this works at different levels.

Student Collaboration for Learning

One aspect of student collaboration is to learn. Although students do need individual time to process information, they need time to discuss their understanding. In order to do this effectively, students need guidelines and expectations. They need to be explicitly taught the skill of collaboration. They must receive timely feedback so they can improve. Many of the strategies mentioned thus far are supportive of student collaboration. The key is modeling these strategies for and with students repeatedly until they can do it independently. Make sure you provide feedback to students so that they can adjust. Table 5.1 is an example of a document you can put on a clipboard and reference as you meet with students.

Student Collaboration for Improving Work

When completing a PBL unit, students work intensely on the creation of a product. For kindergarteners and first graders, it is typically only a group product that they create while third grade, and beyond, completes a group and individual product. These products are critiqued by students as a way to improve student work. Critiquing is a process in which students get specific help that improves their products. This process is often best implemented by

TABLE 5.1

Feedback tool

Collaboration Feedback*

Below are characteristics of collaborative learning along with a possible feedback statement to share with students.

- ❑ Plan and make group decisions.
  - ◼ You used one of the classroom norms. That is a great way to work with others to make decisions
- ❑ Communicate about thinking with the group.
  - ◼ You did a good job using words from the word wall
- ❑ Asks for help.
  - ◼ It was great that you asked for help from your classmate
- ❑ Helps another person with questions and explanations
  - ◼ I like the way you asked them what they already know to help them
- ❑ Keeps others on track
  - ◼ It was very nice how you got your classmate to get back to finishing the task
- ❑ Offers words of encouragement
  - ◼ I heard you tell your classmate that they did a good job. Thanks for encouraging them

---

* The characteristics are compiled from various sources

using protocols as mentioned in Strategy 2. A really easy protocol to use with students is a gallery walk. Students use sticky notes to write or paper with short statements and pictures to mark and leave the feedback on the students' work.

Student Collaboration for Creating Work

Having students work together for the purpose of creating one product is often the hardest collaboration of all. It requires you to really have a clear picture of how you will have students work on the creation of the product. Here are some helpful questions to think about before you gather structures and resources that will support your goal:

- ❑ Does everyone have to do the same amount of work?
- ❑ Does everyone have to contribute to the dialogue equally?
- ❑ Are we having students take on roles?

Once you have the answers to these questions, then you must structure the time students work on the product with the answers in mind. For example, I may want everyone to contribute when they talk. I want students to use their strengths and this may mean they don't do the same amount of work. I also don't want to use roles to not confuse them with the duties they already have as a class. This would mean I would be really sure I have a way to make sure there are lots of opportunities for students to share their thinking and for me to capture that thinking.

Designing Lessons Using the Key Strategies

As stated at the beginning of this chapter, this section explains how to plan lessons in a project using some of the key strategies. It is an elaboration of building inquiry explained in Chapter 3.

Identify Key Deadlines

The inquiry you design needs to be through the lens of key deadlines. Key deadlines include school requirements, final products and major milestones such as experts, significant reflections or critique revision items. You can use a desk calendar or a digital calendar to set the dates. It is helpful to color code the different deadlines. For example, items in red might be the final products, blue items might be school-required assessments such as Common Assessments, and purple might be for times where experts, reflections and/or critique and revision activities occur. It is more efficient and effective to have significant reflections completed when you have critiques or they have a presentation by an expert. Significant reflections are specific times students should reflect on the content or process occurring in the project that purposefully helps them move forward in their thinking.

Create a List of Strategies or Activities That Need to Occur to Reach the Deadline

This step is crucial for student success in your PBL unit. Deadlines should be firm. You should never feel forced to move deadlines, and you won't if students are given proper preparation to meet them. Strategies or activities must align with the deadline; therefore, it is important to think about everything students need to meet the key deadlines. This can be done in any written form, such as a list or an outline. The items do not have to be put in a specific order.

Before you make a list of strategies or activities to support a deadline, you need to think about what students need to know or do. This helps you align the activities to their proper learning objectives. Try working with a group of people to think of several possible activities. This enables you to select the best activity or activities that meet your students' learning needs.

Write the Lesson Plans Using Appropriate Strategies

This is probably the hardest step. This is because you must plan with fluidity in mind. You need to write plans in a way that meets all of these requirements: aligns with district or school expectations; can go in a different order so that students can drive the learning; and have flexibility for things to not go as planned. Deadlines need to be moved as little as possible to ensure the PBL does not get extended beyond a hard date that must not be crossed. However, the lessons themselves need to be fluid and responsive to the students in the room. Make sure you include the strategies mentioned above, especially the staple strategies.

References

"The 5E Instructional Model." *nasa.gov*, n.d., https://nasaeclips.arc.nasa.gov/teachertoolbox/the5e.

"Collection of 3-ACT Lessons." *G Fletchy.com*, n.d., https://gfletchy.com/3-act-lessons/.

Edutopia. "How to Teach Math as a Social Activity." *How to Teach Math as a Social Activity*, n.d., https://www.edutopia.org/video/how-teach-math-social-activity.

EL Education. "Improving Teachers' and Students' Conceptual Understanding of Mathematics." *Improving Teachers' and Students' Conceptual Understanding of Mathematics*, n.d., https://eleducation.org/resources/using-a-problem-based-task-with-fourth-graders-to-create-deep-engagement-in-math.

EL Education, director. *Using a Problem-Based Task With Fourth Graders to Create Deep Engagement in Math*, n.d., https://vimeo.com/117861347.

Houghton Mifflin Harcourt. "Math Workshop Video." *Math Workshop Videos*, n.d., https://mathsolutions.com/math-workshop-videos/.

Laying the Groundwork: 5 Practices for Orchestrating Productive Mathematics Discussions. Youtube, n.d., https://www.youtube.com/watch?v=Ad8gcL89V3c.

Liljedahl, Peter. *Building Thinking Classrooms in Mathematics.* Corwin Mathematics, 2021.

National Council of Teachers of Mathematics. *The 5 Practices for Orchestrating Productive Mathematical DIscussion.* NCTM, 2011.

"The National Paideia Center." *paideia.org*, n.d., https://www.paideia.org/.

PBL Works. "Using a Project Wall to Support Gold Standard Project Based Teaching." *PBL Works.org*, n.d., https://www.pblworks.org/blog/using-project-wall-support-gold-standard-project-based-teaching.

"Project Zero's Thinking Routine Toolbox." *PZ's Thinking Routine Toolbox*, n.d., http://www.pz.harvard.edu/thinking-routines.

"Protocols." *School Reform Initiative.org*, n.d., https://www.schoolreforminitiative.org/protocols/.

Right Question Institute. "What is the QFT?" *RightQuestion.org*, n.d., https://rightquestion.org/what-is-the-qft/.

Ritchhart, Ron, and Mark Church. *The Power of Making Thinking Visible*. B Jossey-Bass, 2020.

"Teacher Resources and Classroom Supplies." *Hand 2 Mind*, n.d., https://www.hand2mind.com/.

Texas Instruments. "Building Concepts." *Texas Instruments Building Concepts in Mathematics*, n.d., https://education.ti.com/en/building-concepts.

"Welcome to the Mathematics Assessment Project." *Mathematics Assessment Project*, n.d., https://www.map.mathshell.org/.

INQUIRY-BASED TASKS AND PBL UNIT EXAMPLES

CHAPTER 6

Complete PBL Unit Example

This chapter features a complete example of a second grade PBL unit, including a calendar, sample student handouts and sample assessments. This example, created by Chris, shows how a project can have math drive the learning of other subjects as well as itself. This project was inspired by a project Chris did with his students a few years ago. Students' interest in pirates is at the heart of this idea. Handouts for the project can be found in the Appendix in the Complete PBL Unit Example Handouts section and www.routledge.com/9781646322114. The project is outlined as follows:

- ❏ Subject/grade level
- ❏ Standards
- ❏ Math/science knowledge for teachers
- ❏ Time frame
- ❏ PBL question
- ❏ Situation/problem
- ❏ Materials/resources
- ❏ Possible PBL unit introduction
- ❏ Anticipated questions
- ❏ Schedule/calendar
- ❏ Formative assessments
- ❏ Reflection prompts
- ❏ Summative assessments

DOI: 10.4324/9781003237358-9

PBL UNIT EXAMPLE: PIRATE TREASURE

Subject/Grade Level

Second Grade

Standards

[Note: While planning, some standards might be a preview or they might be a review depending upon the time of year this is completed]

Common Core State Standards & Mathematical Practices – Math

- ❏ 2.NBT.B.7: Add and subtract within 1000, using concrete models or drawings and strategies based on place value, properties of operations, and/or the relationship between addition and subtraction; relate the strategy to a written method. Understand that in adding and subtracting three-digit numbers, one adds or subtracts hundreds and hundreds, tens and tens, ones and ones; and sometimes it is necessary to compose or decompose tens or hundreds.
- ❏ 2.MD.A.1: Measure the length of an object by selecting and using appropriate tools such as rulers, yardsticks, meter sticks and measuring tapes.
- ❏ 2.MD.B.5: Use addition and subtraction within 100 to solve word problems involving lengths that are given in the same units, e.g., by using drawings (such as drawings of rulers) and equations with a symbol for the unknown number to represent the problem.
- ❏ 2.MD.D.9: Generate measurement data by measuring lengths of several objects to the nearest whole unit, or by making repeated measurements of the same object. Show the measurements by making a line plot, where the horizontal scale is marked off in whole-number units.
- ❏ MP1 Make sense of problems and persevere in solving them.
- ❏ MP4 Model with mathematics.
- ❏ MP5 Use appropriate tools strategically.

Next Generation Science Standards

- ❏ K-2-ETS1-2: Develop a simple sketch, drawing or physical model to illustrate how the shape of an object helps it function as needed to solve a given problem.
- ❏ K-2-ETS1-3 : Analyze data from tests of two objects designed to solve the same problem to compare the strengths and weaknesses of how each performs.

Common Core State Standards-ELA/LIT

- ❏ RI.2.1: Ask and answer such questions as *who, what, where, when, why* and *how* to demonstrate understanding of key details in a text.
- ❏ W.2.8: Recall information from experiences or gather information from provided sources to answer a question.
- ❏ SL.2.4: Tell a story or recount an experience with appropriate facts and relevant, descriptive details, speaking audibly in coherent sentences.

Math/Science Knowledge for Teacher

Math/Science Connected to Project Standards

Students will need to know how to count numbers up to 1000. Each group will have between 100 and 200 pieces of treasure and each person will have between 25 and 50 pieces. They will be able to add and subtract amounts of treasure as they measure the weight, in ounces, while determining how many pieces of treasure will fit on their boat.

Students will also be using 1-inch cube blocks or rulers to measure the dimensions of their boats. Because the volume of the boat will be needed, students will need to add the number of cubes that can fit into their boats or to use technology to multiply the 3 boat dimensions (LxWxH).

Math/Science Facts

The students will be exposed to ideas and concepts that they will be seeing throughout their years with math and science. For example, they will see an equation for density. Buoyancy and density are related. Buoyancy is not only a fun spelling word, but it's all about floating in water. And it is just the idea that as long as the density of the object in the water is less than the density of water, then it will float. Typically, the density of water is a little less than one ounce for every two cubic inches. So, if the weight of their boat (plus their treasure) is known (in ounces) and the volume of their boat is known (in in^3),

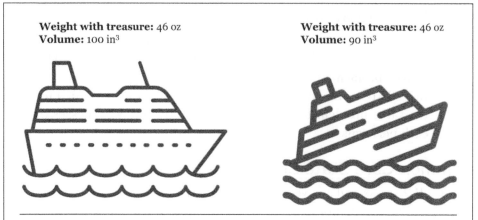

Weight with treasure: 46 oz
Volume: 100 in³

Weight with treasure: 46 oz
Volume: 90 in³

Figure 6.1 Visual representation of buoyancy. Images are courtesy of istockphoto.com

then an approximate value for buoyancy will be known. If the volume number is more than twice the weight number, then their boat should float. Here's an example (see Figure 6.1): if the weight of a boat (with its treasure) is 46 ounces, then as long as the volume of the boat is greater than 92 cubic inches, then it should float.

There are many different ways you could scaffold for these values. The easiest is if the number for the volume is larger than twice the number of inch cubes used for the volume, then it will float. But, with calculators, you could go much further with the calculations. You will have to estimate the "volume" of the boat. If the boats are rectangular prisms (boxes) then it is simply length times width times height (L x W x H) to get the volume. This is a great activity to use for predictions as well, e.g., how much treasure can they put in their boat before it sinks? If you are using pennies for the treasure, then 12 pennies weigh about 1 ounce. That means 100 pennies would be around 8 ounces and 16 cubic inches works for a rectangular prism that is 4 inches by 4 inches and 1 inch tall.

Time Frame

10 to 15 hours

PBL Question

How do we, as pirates, carry our treasure safely? Alternate: how do ship loaders get their goods to places safely?

Situation/Problem

Make-believe is so much a part of a second grader's mindset. They can spend hours pretending to be superheroes and villains. Many are familiar with pirates and their life on a ship. In this project, students take on the role of a pirate ship loader. Students read a book about pirates. They are given maps of a certain area and they are instructed to explore the area to find buried treasure. Once the students have found the treasure, they will be instructed to design and build a boat to sail to a new land and take the treasure to bury it there. When they get to their new land the students will create a map showing where they have taken the treasure with written instructions to tell how to find the treasure they have buried there.

Materials/Resources

- ❑ Each team will need two sheets of aluminum foil (one for practice and one for the final build). Each sheet should be the same size for each team. For example, a 10 inch by 10 in square piece. In addition, there should be enough aluminum foil for each student to practice building a boat.
- ❑ Each team will be given enough treasure to guarantee that if all of the treasure is placed on the built boats, then the boat will sink. In addition, there should be enough treasure for the total in the class to be in the hundreds so that students can practice addition and subtraction with the treasure. Pennies are an easy thing to use.
- ❑ Each team will be given paper to draw treasure maps along with paper to write out the instructions for finding the buried treasure.
- ❑ Each student will be given a worksheet to record their boat's "volume" and the amount of treasure the boat can hold ("weight").

❏ At least one scale for students to measure the weight of the treasure and the weight of the aluminum foil.

❏ Rulers to measure their boats and distances on their treasure maps. These distances should be done in inches.

❏ Inch cubes for measuring. These can be used for measuring length as well as volume.

❏ Calculators or technology to do buoyancy calculations, if desired.

❏ Large tub or kiddie pool to float the boats.

❏ Possible books to read:

1. *Pirate Potty* by Samantha Berger
2. *Ghost Pirate Treasure* by Geronimo Stilton
3. *Pirate Spacecat Attack* by Geronimo Stilton
4. *Shipwreck on the Pirate Islands* by Geronimo Stilton
5. *There was an Old Pirate Who Swallowed a Map* by Lucille Colandro
6. *Thea Stilton and the Tropical Treasure* by Thea Stilton

Possible PBL Unit Introduction

Students will read a pirate book and they will be shown a map of where "treasure is buried." They will be told that the "treasure" is hidden in the classroom and they must find it. Once they have the treasure, they will need to build a boat to sail the treasure to a new land. Finally, they will have to draw a treasure map that explains where they have "buried" their treasure in this new land.

Anticipated Questions:

❏ What materials will we use to build the boat?

❏ How much treasure can our boat carry?

❏ Will we have to carry all of the treasure in our boat?

❏ How much does the treasure weigh?

❏ How big are things on the treasure map?

❏ Were pirates real?

❏ Are there pirates today?

- ❏ Where do the pirates live?
- ❏ Do heavy things sink and light things float?
- ❏ How are maps created?

Schedule/Calendar

[See Chapter 3 "PBL Elementary Schedule" for further guidance]

Week 1

Day 1 [Project Time: 1.5 to 2 Hours]
Project Details Overview:
- ❏ Get students engaged with the project with a kickoff
- ❏ Create teams and gather knows/need to knows
- ❏ Discuss first steps and deadlines
- ❏ Have a reflection time
- ❏ Journal entry

Place students in teams of four and hand each team a "treasure map" of the classroom or other area where the treasure is hidden. Have each team's treasure hidden in a different place and labeled. This keeps other teams from taking another team's treasure. Discuss how to read the map and what things they know about the map. Allow students to search to find their treasure and then have them bring their treasure back to their sitting area.

Ask students to think about how they could determine how much treasure there is for the whole class. This leads to a discussion of how pirates carried their treasure on their ships. Students are told they will have to build a boat to carry their treasure to a new place where no one can find it. They are told that they won't be able to carry all of their treasure because there is too much for the boats they will make. But their goal is to carry as much of the classes treasure on the boats the teams will be making. They will have to calculate how much of the treasure their boat can carry and the total amount that can be carried by all of the boats.

Collect knows and need-to-knows (K/NTKs). Discuss what they should work on the next day to help them prepare to write in their journal. Explain that on Day 6, also known as Sink Day, they will be seeing how much treasure their boats will carry and they will get to add treasure until their boat sinks. Possible reflection questions for today:
- ❏ What are you excited about with this project?
- ❏ How much of your treasure do you think your boat will carry?
- ❏ Do you think the class will be able to put all of the treasure on all of the boats that are created? Why or why not?
- ❏ What special skill do you bring to your team?

Students close out with writing in their journal. They share what they believe should be worked on the next day.

Week 1, *continued*

Day 2 [Project Time: 1 Hours]

Project Details Overview:

- ❏ Review K/NTKs. Add additional things students want to know
- ❏ Team building: Name team and their boat
- ❏ Practice building a rectangular prism (rough draft of boat)
- ❏ Introduce vocabulary words: rectangular prism, buoyancy and volume [note: these are not required vocabulary terms for this grade level but they will be interacting with these words in future learning situations]

Review the K/NTK list as a class. Add additional items to the list. Teams come up with a team name and name for their pirate ship.

Students are told how they will make a model of a boat. Students are introduced to the vocabulary word for their word wall, "rectangular prism." Each student is given a sheet of aluminum foil. Students practice folding the foil to make a rectangular prism.

Students are asked if a rectangular prism could be used to create a model of a boat? Students discuss how they might make sure a boat floats. They are given a new vocabulary word "buoyancy," and they learn about the concept of buoyancy and the ability to float. Students are given another vocabulary word, "volume." They are told that if the volume of an object is bigger than twice an object's weight, then the object will float in water. Students discuss times that objects have floated in water. Students are asked to think about how they might guarantee a boat floats when it has treasure in it. Buoyancy and Volume are added to the word wall.

Special Note: have each student put their name on their boat because these boats will be used later.

Day 3 [Project Time: 1.5 to 2 Hours]

Project Details Overview:

- ❏ Interact with K/NTKs, adding or lining through items as needed
- ❏ Practice measuring weight with a balance beam or digital scale
- ❏ Practice measuring items using a cm ruler
- ❏ Practice adding and subtracting using each team's treasure (note: this is a heavy addition/subtraction day)
- ❏ Journal entry

Review the K/NTK list as a class. Add additional items or remove items to the list. Look at the word wall and review the meaning of each of the words: rectangular prism, buoyancy and volume. Hand out the Volume and Weight Record Worksheet to each student (to be completed on Day 3 and 4).

Teams will learn how to measure weight and they will use inch cubes or a ruler to measure the dimensions of the rectangular prisms they made Day 2. They will help each other with the measuring using a balance beam and a ruler. Each person will record the amount of treasure (number of pennies) on the scale and the weight that they read. Then they will add pennies and practice using their academic vocabulary: "I added _____ pieces of treasure to the amount on the scale and now I have _____ pieces of treasure on the scale." They will also do this by removing pennies: "I subtracted _____ pieces of treasure from the scale and now I have _____ pieces of treasure on the scale." The same will be done for the weight: "I added _____ pieces of treasure to the amount of ounces on the scale and now I have _____ ounces of treasure on the scale." And, with removing pennies: "I subtracted _____ pieces of treasure from the scale and now I have _____ ounces of treasure on the scale."

Week 1, *continued*
The goal is to have a record of at least five different amounts of pennies and the weight of that many pennies. These values will be recorded on the worksheet. They will end the class with a journal entry about the kind of treasure they would carry on their boat.

Day 4 [Project Time: 1.5 to 2 Hours]

Project Details Overview:

❑ Interact with K/NTKs, adding or lining through items as needed
❑ Practice measuring weight and length
❑ Introduce "volume" and calculate volume of a rough draft model
❑ Complete Volume and Weight Record Worksheet
❑ Draw pictures of their final boat build and write improvements from their rough draft [additional: label with dimensions]

Review the K/NTK list as a class. Add additional items or remove items to the list. Students will be introduced to the word "cubic inches" and they will add it to their word wall. Review the directions for the Volume and Weight Record Worksheet. They will take the measurements of their rough draft boats and calculate the volume of their boats by counting the inch cubes needed to fill the boat, or by using technology such as a calculator. Team members will help each other verify the values. All measurements will be done in inches and the volume will be in cubic inches. When all team members are finished with their calculations, they will record the volume with the data they collected on the amount of treasure and the weight of that amount on their worksheet.

Each person then will draw a picture of their boat. You may want them to label the dimensions and write the volume on the drawing. If you do, determine whether you need to teach them how to label dimensions. Team members will ensure that each person's volume numbers are accurate. Once each member of the team has completed their Volume and Weight Record Worksheet, then team members are to improve their boat drawings by adding at least one thing that makes their boat unique.

Each team will discuss which of the team member's boats looks the strongest; which boat will carry the most treasure; and which team member is the best builder. Each team member will write down ways they could improve their boat to make it better looking and to carry more treasure. Students should start looking at their peer's boats and start thinking how much treasure the class will be able to carry onto the boats.

Day 5 [Project Time: 1 to 1.5 Hours]

Project Details Overview:

❑ Review K/NTKs
❑ Introduce the idea of density and buoyancy [note: this is a science heavy day]
❑ Students explore the volume/weight ratio
❑ Discuss size vs. weight: does a "bigger" boat have to weigh more?
❑ Students determine where their hiding place is and put written instructions of how to get to it in their journal

Week 1, *continued*

Review the K/NTK list as a class. Add additional items or remove items to the list. Students will review weight and volume and will look at the ratio of the two. This leads to the concept of density and buoyancy or when the volume (in cubic inches) is more than twice the weight (in ounces). Show examples of objects where: (1) there is a small volume and a large weight, (2) there is a large volume and a small weight and (3) it is hard to tell just by looking at the object whether it will be buoyant or not. Students discuss the question: "Does a bigger boat have to weigh more?" Good example would be one of their rough draft boats made from aluminum foil and a small box that is much heavier. Once this activity is complete, the students will start thinking about where they want to hide their treasure.

During social studies or literacy blocks or station rotations, build in time for students to write and explore the map skills and connect the concept of creating the new map back to the Entry Event. The goal is to have each student write directions on how to get from their desk to where the treasure will be hidden.

In the writing workshop, focus on being descriptive. The directions will be submitted in their journal entry. Remind students to be thinking about how they can carry all of the treasure on the boats being created in each group.

Week 2

Day 6 [Sink Day] [Project Time: 2 Hours]
Project Details Overview:
- ❏ Review K/NTKs
- ❏ Practice measuring by using rough draft boat and treasure
- ❏ Students add treasure to their rough draft boats until they sink
- ❏ Students create tables with two columns: Amount of Treasure and Did Boat Sink? (Y/N)
- ❏ Data table is written into journal

Review the K/NTK list as a class. Add additional items or remove items to the list. Today is the first Sink Day. After today, students will have opportunities to see how much their boat designs will carry until the boat sinks each day. You will need to have a kiddie pool or deep sink available to place the boats into for sinking each day. Students start by taking their original boats or they can build new boats. If they build a new boat, they will have to measure the dimensions and they will have to calculate the volume of this boat.

Review density and buoyancy and ask each student to predict how much treasure (ounces) they can add to their boat before it sinks. Have students measure that amount (using a scale) and wait for their turn to try and sink their boat. Each person will get to add treasure to their boat to see if it sinks or floats with their predicted amount. Then they will add more treasure (one at a time) until the boat sinks. Record the actual number and record the total for the class once every boat has been tested.

Have the students put this information into a table in their journal and say if their prediction was correct or not.

Week 2, *continued*

Day 7 [Project Time: 2 Hours plus any intervention time]
Project Details Overview:
❏ Review K/NTKs
❏ Formative assessment on adding and subtracting numbers less than 1000
❏ Formative assessment on measuring weight
❏ Work with those needing additional help with adding/subtracting
❏ Work with those needing additional help with measuring weight
❏ Redesign the boat

Review the K/NTK list as a class. Add additional items or remove items to the list. It is time to do some intentional assessment of the math skills of adding and subtracting numbers to 1000 and the science skill of measuring weight. This is a traditional assessment that helps inform the refinement of boat building revision that occurs the next day. Using an electronic assessment, such as Google Forms or Quizzizz.com will give you results quickly so that you can provide small group instruction to those who need additional help with adding/subtracting or measuring weight.

After the assessment, students can plan how they refine their boat individually or with their teams. Make sure they understand that it is alright for every member of the team to build a different style of boat today. The goal is to have just one boat that represents the team for the presentation day and the goal for the class is for all of the treasure to be carried by the boats created by the teams.

Teams should start thinking about their presentations that will be done on Day 11. See Days 10 and 11 for specific requirements for each group.

Day 8 [Revised Boat Building Day] [Project Time: 1.5 to 2 Hours]
Project Details Overview:
❏ Review K/NTKs
❏ Read a book about ships
❏ Class creates criteria for what a "great ship" looks like
❏ Students build their final boat design. Each person builds a boat and the team will test all of the boats in their team to determine their "best" boat
❏ Students write a rough draft story about an adventure on their boat

Review the K/NTK list as a class. Add additional items or remove items to the list. Read a book about ships that really help students identify the qualities of a ship. Create a list of what a "great ship" looks like and post it in the classroom. Each student will build their boat. It is possible that the students in a team will build the same boat but encourage each person to create a boat that looks different. Remind them they will all get to test their boats and determine which boat is the "best" boat in their teams. The team goal is to have just one boat for each team that represents the team for the presentation day. The goal for the class is for all of the treasure to be carried by the team boats.

Each student will do a quick write about an adventure on their boat during Writer's Workshop. The story will be peer reviewed on Day 9. Teams continue to prepare for their presentation on Day 11.

See Days 10 and 11 for specific requirements for each group.

Week 2, *continued*

Day 9 [Project Time: 2 Hours]

Project Details Overview:

❏ Review K/NTKs

❏ Peer Review of rough draft stories

❏ Finalize artistic elements to boats

❏ Teams select the boat that will represent their team

❏ Final collection of measurements for the team's boat

❏ Predictions of how much treasure the team's boat will carry

❏ Create a class table of each team's boat and the predicted amount of treasure for each boat

❏ Students write about how they are feeling about their boat in their journal

Review the K/NTK list as a class. Add additional items or remove items to the list. Students peer review (see Chapter 5 for ideas) their written rough drafts about their boat adventure. When they finish the peer review, they make final touches to their boats to enhance their aesthetics. Even though each team will select one boat to represent the team, every boat will get tested to see how much treasure it can carry. Tell students to recheck their boat's weight after they have added things to the boat since that will affect how much treasure the boat can carry. Have them predict the new amount of treasure based upon this new boat weight.

Once each team has selected their boat, then the dimensions will be retaken and a prediction of how much treasure the boat can carry will be made. Each team will report on why they chose the team's boat and the amount of treasure their team boat can carry. A class table will be created with that information from each team. The goal is to have just one boat that represents the team for the presentation day and the goal for the class is for all of the treasure to be carried by the boats created by the teams. In the class table, have the amount of treasure for each team's boat and have the class add all of the amounts to see if it adds up to the entire amount of treasure.

As a close, students will write about their boat and their teammates' boats in their journal. Possible prompts: Was your team's boat able to carry all of your team's treasure? What might you have done to improve your own boat and your team's boat to help it carry more treasure?

Day 10 [Boat Presentation Prep Day] [Project Time: 1 to 1.5 Hours]

Project Details Overview:

❏ Review K/NTKs

❏ Teams plan and practice a presentation

❏ Formative assessment on adding/subtracting

❏ Formative assessment on measuring using a ruler and a scale

❏ Formative assessment on calculating density (buoyancy)

Review the K/NTK list as a class. Make sure all questions are addressed that are essential for the presentation day.

Teams plan and practice their presentations. List each of the key components for the presentation in a prominent location:

Week 2, *continued*
(1) The name of their team, boat and why that boat was selected (2) The dimensions of their boat (3) The volume of their boat (4) The predicted amount of treasure their boat will carry (5) A short story (with each team member telling a part) about an adventure they had on their boat At the end of the day, each team will present to another team to get some feedback.

Week 3

Day 11 [Boat Presentation Day/Dress Like a Pirate Day/Sink Day 2]

Project Details Overview:

❏ Have duties for each student to prepare the classroom and to greet parents and guests

❏ Announce the order of presentations

❏ Each team does their presentation and then sinks their boat

❏ Each team's "Actual amount of treasure needed to sink the boat" is recorded

❏ End the presentation and thank visitors

❏ Class discussion on the class table of data – "What do you see?" and "What do you wonder?"

❏ Each student records their responses to the discussion as journal entry

Students will be dressed like pirates. Acknowledge their outfits including those parents/guests that dress up. Make sure the schedule of presentations is written somewhere prominent so that everyone is fully aware of the order of presentations. Make sure the kiddie pool or deep sink area is clear so that most people in the room can see the sinking of the boats. Have a prominent place where you will keep the table of data from the sinking process. Review duties with all students who will be helping receive the guests or who will escort the guests out from the presentations.

Do a dry run before the guests show up. Coordinate times so that either you complete just prior to the lunch period or just prior to the end of the day.

Students complete the presentations in front of guests. Thank the guests for their participation and feedback to the students.

After the guests leave, complete a class discussion about the table of data using notice and wonder strategy (What do you see? What do you wonder?). Students share out and record responses in their journal.

Day 12 [Reflection Day and Summative Assessments]

Project Details Overview:

❏ Summative Assessment (if needed)

❏ Review K/NTKs and discuss with class – have all of the NTKs been answered?

❏ Conduct a written summative (if required by your school) that includes adding/subtracting numbers to 1000 and measuring with a ruler and a scale. Have students calculate whether a certain box, with given dimensions and weight, would sink or float

❏ Written reflection

Week 3, *continued*
Start the day by celebrating the successful completion of the project. Have students share the best parts of the presentation day.
If your school requires a summative assessment (test or quiz) to show growth/progress, then give this to students early in the day separate from the celebration and with enough time to complete.
Take time to look, for the final time, at the K/NTKs and do some reflecting – did every NTK get answered?
Have students write a final time in their journal about their work as a team.

Formative Assessments

The following is a brief description of the formative assessments that occurred in this project. If you implement this project in your classroom, adapt it to your school context and needs.

- ❑ **Know/Need to Know list:** this list is a way for you to see how students make sense and persevere in problem solving which is a mathematical practice. It is a visible representation of the students' understanding over time.
- ❑ **Entrance or exit tickets:** entrance and exit tickets are short tasks completed at the beginning or end of class. When completed at the beginning, it enables them to focus on the day. When completed at the end, it allows for them to show their understanding for the day.
- ❑ **Quizzes:** traditional assessments can always be used in a project. A quiz could be given over the target math standards based on your district's scope and sequence. For example, if measurement is the focus, build a quiz around the measurement skills students learned. Or, you might have a quiz on adding and subtracting numbers to 1000 before students get to the sink day in the project. It is also a good idea to do a quiz (see an example on mathpbl.com) where multiple learning targets are combined like in the project.
- ❑ **Whole-class discussion and observation:** each day of the project, there are whole and small group interactions. These are opportunities to check for understanding by asking students questions. It is best practice to have a list of questions you are wanting to ask students during each interaction. The goal is to inform you of what students know and don't know.

Reflection Prompts

The following are reflection questions that can be used during the project. You do not need to use all of the questions. You can pick the questions that are applicable to your students. They can be given as exit tickets or answered in a journal. It is always important for students to think about what they did in class each day. It is always a good idea to mix questions between content, social and emotional learning questions and product type of questions.

- ❏ What tasks do you and your team need to complete in class tomorrow?
- ❏ Tell a story about loading your boat with treasure. What kind of treasure would you carry on your boat?
- ❏ What have you learned about floating in the water today?
- ❏ Was your prediction good? Why or why not?
- ❏ When you measure with your ruler, what do the little lines in between the main numbers mean?
- ❏ How are you feeling about your boat and teammates' boats? Will your team's boat carry the amount of treasure you predict? Why? Will your team's boat carry the most in the class? Why? Is their team's boat the best looking? Why?
- ❏ Which of your team partners have you helped today, or which of your team partners have helped you today? What did that help look like?
- ❏ What would you ask a pirate if one came to our class tomorrow?
- ❏ Were you a good team member? What things did you do that really helped your team? Who in your team, besides yourself, was the person who worked the hardest? What things did that person do that really helped your team?

Summative Assessments

The following is a brief description of the summative assessments that occurred in this project. If you implement this project in your classroom, adapt it to your school context and needs. For example, a formal written test is often given after the reflection day. You can still do this in PBL. Here is the summative assessments:

- ❏ **Presentation:** this is a product that every student must complete and revise at least twice within the project. At least two revisions help ensure students produce a quality product. You must scaffold through activities that help with the creation of the product. For this project students are building boats and they are making predictions about the amount of treasure that their boat can safely carry.

CHAPTER 7

Inquiry-Based Task

Examples

This chapter includes examples of inquiry-based tasks and units for grades K–2. A task is a condensed form of a PBL unit, similar to a robust performance assessment or an extended activity. Tasks can vary in length of time to complete, ranging from a couple of math blocks to a few days. Tasks are great for covering standards over just a few math blocks. They are also a great gateway into completing a full project, if you are feeling a bit anxious about making the jump to implementing a full Project-Based Learning unit. The tasks do not include complete details like a textbook. The main details are given to help you implement the task. You can find additional resources for each task at mathpbl .com.

In this chapter there are three tasks; one for each grade level (K–2). After the third task there is a unit that is built from one of the first tasks to demonstrate how you can take a strong idea and you can extend it to a full PBL unit.

- ❏ Title
- ❏ Subject/grade level
- ❏ Common Core State Standards and mathematical practices
- ❏ Time frame
- ❏ Task question

DOI: 10.4324/9781003237358-10

❑ Situation/problem
❑ Materials/resources
❑ Possible misconceptions/expected nswers
❑ Example lesson: mini-lesson (whole class), work time (individual or small groups), debrief (whole class)

The tasks in this chapter are divided by grade level. However, the idea can be done at any grade level. Take the time to make sure you look at each task and feel free to adapt to fit the grade level you teach. The tasks are:

❑ Task 1: Shapes and Visuals (Grade K)
❑ Task 2: Measuring (Grade 1)
❑ Task 3: Puzzle Boxes (Grade 2)

TASK 1: SHAPES (KINDERGARTEN)

Subject/Grade Level

Shapes & Visuals, Kindergarten

Common Core State Standards and Mathematical Practices

- ❏ K.G.A.2 Correctly name shapes regardless of their orientation or overall size.
- ❏ K.G.B.5 Model shapes in the world by building shapes from components (e.g., sticks and clay balls) and drawing shapes.
- ❏ MP6 Attend to precision.
- ❏ SL.K.5. Add drawings or other visual displays to descriptions as desired to provide additional detail.

Time Frame

3 to 5 hours, depending upon the extent of the skyline.

Task Question

How can we create a skyline representation of a city using shapes?

Situation/Problem

Students have been learning about the shapes around them. For this task the students are given one of ten famous city skylines. They will create a drawing of that skyline using geometric shapes including rectangles, triangles, circles, squares, or hexagons. The drawing will include a key that explains the geometric shapes used to make the drawing. When they have completed their skylines, they will tell the name of the city and where it is located. They will describe to another classmate the skyline using academic language for the geometric shapes that were used.

Materials/Resources

- ❏ Classroom set of wooden blocks or other types of blocks.
- ❏ Tracing paper.
- ❏ Construction paper, tracing paper, chart paper, butcher paper or other paper as needed.
- ❏ Rulers or other straight-edge objects for drawing straight lines.
- ❏ Adhesive to attach shapes to the final paper.

Possible Misconceptions/Expected Answers

There are many combinations of shapes that create a city skyline. We want the students to know the shapes of squares, rectangles, triangles, circles and hexagons. But what do you call something that is a combination of a triangle and a rectangle? Students might want to just classify the entire composite as either a rectangle or a triangle. Students will need to be able to break a composite figure into its smaller shapes before describing them.

Students should be able to point to their skyline and tell a classmate what shapes make up the skyline. For example, "The top of this building is a triangle and below it is a rectangle with windows that are also rectangles. The bottom of the building is a large rectangle and there are circles (here) and rectangles (here) that are windows and doors."

Example Lesson

Mini-Lesson (Whole Class)

1. Start with a quick review of shapes. This can be done simply by showing a shape and asking students to state the name of the shape or the teacher saying the name of the shape and having students find that shape in their classroom. The students are then shown pictures of buildings in their town or city. They are asked to name the shapes they see in the pictures such as "the window is a square and that door is a rectangle."

2. Each table group has blocks and they are asked to see if they can replicate the picture using the blocks on their tables. Have each student tell you what they are building and what shapes they are using to build their copy of the picture. Have multiple copies of pictures on each table so that students have the option of building more than one building with their blocks.

3. After ten minutes, volunteers present their buildings using their academic language of shapes.

Work Time (In Pairs)

1. For the first 10 minutes, students take copies of their selected skyline and they use tracing paper to trace the outline of the skyline. This will become the base for making a large sheet of paper that will be used for attaching shapes to replicate the overall skyline later in this task.

2. For the next 15 minutes, students work in pairs with their tracings while they discuss what shapes they see in the traced skyline. If they need to make the shapes more visible, they can draw in the rest of the shape not represented by the skyline tracing.

3. For the last 15 minutes pairs of students are put together to create groups of four students. One pair then explains their skyline and what shapes they noticed in their tracings. Emphasize good listening for the two students hearing from the explanation. Once the first pair has completed then the second pair explains their skyline tracing.

Debrief (Whole Class)

1. Ask students to look at their tracings and the shapes they have decided will represent their skyline. Emphasize that today we focused on the shapes that will make the highest parts of the skyline. Tell students that they will be drawing a large city skyline on chart paper or butcher paper. Then ask students to think about what other shapes they will need besides the shapes for the high parts. Have groups share the other shapes that they will need to complete their skyline drawing.

TASK 2: MEASURING (FIRST GRADE)

Subject/Grade Level

Measurement & Writing, First Grade

Common Core State Standards and Mathematical Practices

- ❑ MD.1.A.2 Express the length of an object as a whole number of length units, by laying multiple copies of a shorter object (the length unit) end to end; understand that the length measurement of an object is the number of same-size length units that span it with no gaps or overlaps.
- ❑ MP2 Reason abstractly and quantitatively.
- ❑ W.1.3 Write narratives in which they recount two or more appropriately. sequenced events, include some details regarding what happened, use temporal words to signal event order, and provide some sense of closure.

Time Frame

1 to 5 hours, depending upon how much time is spent with the writing piece of this task.

Task Question

How do writers use lengths of objects to tell distances in stories?

Situation/Problem

Students are to write a short story about someone who is moving to a new house. The person, animal, or bug who is the narrator must tell their friends where they are going by using a small object (like a paper clip) to describe the distance to their new home. The story must explain how the object was used multiple times to find a total distance from the old house to the new house. Emphasis should be on using an object that is smaller than the main character. For example, if the main character was a guinea pig, then the object might be a paper clip, a straw, or a bean.

Materials/Resources

- ❏ Various objects that can be used to measure desks or tables such as unsharpened pencils, markers, paper clips, erasers or copy paper.
- ❏ Chart paper or whiteboard to record lengths of items in the room and the objects used to measure these.

Possible Misconceptions/Expected Answers

Students are just learning that inches and centimeters are words used when talking about length. They will want to continue using those references even though the point is to use another reference object for measuring. Students will want to know how they can measure using something like a paper clip, a needle, a button or a seed.

The teacher should have some estimated lengths for the desks or tables in their room on a sheet of paper using a few of the objects provided. This will serve as your answer key. For example, a desk might be 12 board erasers long and therefore a student should state that the length is between 11 and 13 erasers long. The teacher should see that students are measuring from the end of one length to the end of the next length such as this:

ERASER 1	ERASER 2

Example Lesson

Mini-Lesson (Whole Class)

1. Students are told great authors have to do research about the characters and other elements of their story. Today, they will be exploring ways to measure the distance between two objects. The students are asked how they would measure the distance from one side of their desk or table to the next desk or table without a formal measurement tool. If any students suggest a ruler, yardstick, or meter stick, remind them how they can't use one of those. As students share, the teacher creates a list of things they say on poster paper or a whiteboard. Once a few students share things like "a shoe," "my arm" or "a piece of paper," then the class should be able to come up with a large number of objects that could be used. Give the students a few minutes to continue to make their own list of objects. Ask the students which of the objects on their list would be the best object to use to do the measurement. Help them

see the importance of measuring using the exact same object over and over. It is not good to just say to use "a shoe" because not every shoe is the same size.

2. In their stories, they will discuss the distances between their character's old home and their new home. Ask students to look at the list of objects they said they could use to measure the distance between desks. Which of these objects could their character use? What other objects might their character use to measure distances? Have each student make a list of objects that their character might use to measure distances.

Work Time (Individually or Pairs)

1. Students are instructed to measure their desk or table using an object from the first list generated in the mini lesson. On a piece of paper, the students draw a rectangle and label it desk or table. Then they write the number of objects needed to find the total length of each side. For example, if their desk was 10 erasers in length they would draw a rectangle, label it "desk," and then write 10 erasers for its length.

2. Students are then each given a paper clip. They are instructed to measure their drawings of their desk (the rectangles labeled "desk.") with the paper clip. Have them write the number of paper clips below where they wrote the previous length. For the previous example they would write 2 paper clips underneath where they wrote "10 erasers." If there isn't an exact number of paper clips (the rectangle is longer than an exact number), then have them use the number of paper clips which is closest. For example, if it was between 2 and 3 paper clips but closer to 2 paper clips, then the student would use 2 as the length.

3. Have students pair up with someone who used a different unit of measure for their initial desk length. Each student will explain how they measured their desk, drawings and will give the number of units they had for a length. Then have each student look at each other's rectangles and the number of paper clips they used for those lengths. Have them discuss which objects were better for measuring the desks and what might be a better object for measuring their rectangles.

Debrief (Whole Class)

1. Student pairs give the lengths of their tables and the objects used to do the measuring. Then they share the lengths of their rectangles along with any objects they found would be better to do the measuring. The teacher creates a table on the board with desk lengths and objects used and the object each group thought was the better object for measuring desks.

2. With the table completed, ask the class to decide which objects would be the best for measuring their desks or tables? Have students explain why they think this is true. Now ask about measuring the drawings of the desk or table. Which object would be best to use and why?

3. Ask students to think about which object would be best for their character to use in their story. They write that object on an exit ticket.

TASK 3: PUZZLE BOXES (SECOND GRADE)

Subject/Grade Level:

Operations & Writing, Second Grade

Common Core State Standards and Mathematical Practices

- ❏ NBT.B.5 Fluently add and subtract within 100 using strategies based on place value, properties of operations, and/or the relationship between addition and subtraction.
- ❏ MP8 Look for and express regularity in repeated reasoning.
- ❏ W.2.2 Write informative/explanatory texts in which they introduce a topic, use facts and definitions to develop points, and provide a concluding statement or section.

Time Frame

1 to 3 hours, depending upon how much time is spent with the writing piece of this task.

Task Question

How do puzzle experts make Number Boxes?

Situation/Problem

Number Boxes are a math puzzle where there are two boxes with numbers in them. The player is asked to switch numbers from one box to the other box so that each box adds up to the same value. Students will create three levels of boxes. The "Easy" boxes will have two numbers in them that add up to no more than 30. The "Medium" boxes will have two or three numbers in them that add up to no more than 100. The "Difficult" boxes will have two to four numbers in them that add up to no more than 150. The class will make and share their Easy and Medium boxes with the first grade classes. They will make Difficult Boxes and share the boxes with the other second and third grade classes. Number

Boxes can just be drawn onto construction paper or they might be drawn and then cut out of a stiff paper. Students will write an "instruction manual" of how Number Boxes work that will include example solutions.

Materials/Resources

❑ Construction paper, card stock or other stiff paper to create the "Number Boxes."

Possible Misconceptions/Expected Answers

Students may look at one of the boxes, and the numbers in it and assume that the total in the other box must add up to the same amount as what is in the first box. Students will see that being able to do mental math for addition and subtraction will help them arrive at the solution quicker.

For the easy boxes there should be mostly single digit numbers since the total must be less than 30. The medium boxes will have a mixture of one- and two-digit numbers and the difficult boxes should only have two-digit numbers.

Example Lesson

Mini-Lesson (Whole Class)

1. Start by asking students if they have ever had a time where they had to share something with a friend or sibling and each person needed to have the exact same amount. Examples might be Halloween Candy or food from a concession stand.
2. Now show the students two groups of single-digit numbers and ask them to think about what they notice about the two groups. What things are the same? What things are different? Possible answers: one group are even numbers and the other group is a mix of odd and even numbers. One group adds up to 16 and the other group adds up to 20 (see Figure 7.1).
3. The students are asked to think about what they could do to have it where both groups of numbers add up to the same amount. They are told that each group has to have the same amount of numbers in them but they can move numbers from one group to the other group. Also let them know that there can be one number two times in a group. For this example, there are 8s in each group and they could be put into the same group if that gives a correct solution.

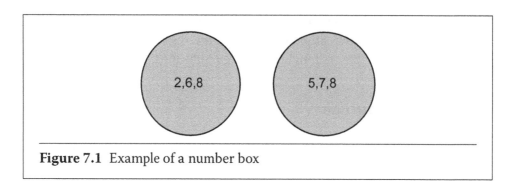

Figure 7.1 Example of a number box

4. Show the students [2, 8, 8] and [5, 6, 7]. Since both groups add up to 18, they are a correct solution to this problem. Ask them to see if there is another correct solution.

5. Now give them two new groups of numbers and ask them to do this again. Start simple with just two or three single-digit numbers in each group. After that is successfully answered, show the class two new groups with a larger value. Allow them to work in pairs for this one. Explain that this is a more difficult problem and that they will be creating groups like this to share with the other classes in first, second and third grades.

Work Time (Individually and Pairs)

1. Each student is told to create five sets of pairs of single-digit numbers where each set adds up to the same value. For example, the sets 6, 4 and 8, 2 both add up to 10. Then have them switch two of the numbers from each group. For this example, we would end up with 6, 8 and 4, 2.

2. Pair students up and have them work each other's five sets of newly arranged numbers. Were any of the sets the same for both students?

3. Working in pairs, have them create five sets of two-digit numbers. Again, have them start by having both groups, in each set, add to the same number. Finally, have the pairs of students get together in groups of four to solve these new sets of numbers.

Debrief (Whole Class)

1. Ask the students if there were any patterns or other interesting things that they discovered in exploring their pairs.

2. Remind students that they will be writing an instruction manual for playing the game and it will have to have some example solutions.

3. Some students may wonder if there is a strategy for finding the solution. This can be given as an extension to find this out or, if you have talked about splitting a group into two equal parts, you could tell them that if you add all of the numbers in the two boxes and then split that number into two equal parts, then that is the amount each box should add up to.

CHAPTER 8

PBL Unit Examples

Project-Based Learning units are designed to last 10 to 15 class hours. The PBL unit examples included in this chapter include some of the same components as the tasks in Chapter 7. They are centered around power standards for particular curricula. The PBL unit examples include the following components:

- ❑ Subject/grade level
- ❑ Common Core State Standards and mathematical practices
- ❑ Time frame
- ❑ PBL question
- ❑ Situation/problem
- ❑ Materials/resources
- ❑ Possible unit introduction
- ❑ Anticipated questions
- ❑ Possible misconceptions/expected answers
- ❑ Formative assessments
- ❑ Reflection prompts
- ❑ Summative assessment
- ❑ Example of a lesson connected to anticipated questions: mini-lesson (whole class), work time (individually or small groups), debrief (whole class)

DOI: 10.4324/9781003237358-11

The PBL unit examples cover the following topics/grade levels:

- ❏ PBL Unit 1: Counting & Cardinality (Kindergarten)
- ❏ PBL Unit 2: Geometry (First Grade)
- ❏ PBL Unit 3: Numbers & Measurement (Second Grade)

PBL UNIT 1: COUNTING AND CARDINALITY (KINDERGARTEN)

Subject/Grade Level

Counting and Cardinality & Opinion, Kindergarten

Common Core State Standards and Mathematical Practices

- ❏ K.CC.A. 1 Count to 100 by ones and by tens.
- ❏ K.CC.A.2 Count forward beginning from a given number within the known sequence (instead of having to begin at 1).
- ❏ K.CC.A.3 Write numbers from 0 to 20. Represent a number of objects with a written numeral 0–20 (with 0 representing a count of no objects).
- ❏ K.CC.B.4 Understand the relationship between numbers and quantities; connect counting to cardinality.
- ❏ K.CC.B.5 Count to answer "how many?" questions about as many as 20 things arranged in a line, a rectangular array, or a circle, or as many as ten things in a scattered configuration; given a number from 1–20, count out that many objects.
- ❏ MP1 Make sense of problems and persevere in solving them.
- ❏ MP4 Model with mathematics.
- ❏ MP6 Attend to precision.
- ❏ W.K.1 Use a combination of drawing, dictating and writing to compose opinion pieces in which they tell a reader the topic or the name of the book they are writing about and state an opinion or preference about the topic or book (e.g., *My favorite book is...*).
- ❏ W.K.5 With guidance and support from adults, respond to questions and suggestions from peers and add details to strengthen writing as needed.
- ❏ W.K.6 With guidance and support from adults, explore a variety of digital tools to produce and publish writing, including in collaboration with peers.

Time Frame

13 to 15 classroom hours

PBL Question

How can we help build up our neighborhood library boxes?

Situation/Problem

Many students have to go far for a book. They hear about neighborhood books. Building neighborhood libraries (https://littlefreelibrary.org/) are growing in popularity. It is just ordinary citizens taking the time to offer free books to people in the neighborhood using a free-standing book house. Students work with the local library or a high school construction class to get the houses and books to put in their neighborhoods. Students also ask the community to donate books. They write short book reviews and number the books for each little library. They must have 20 books each.

Materials/Resources

- ❑ Donated books (at least 60 but preferably 100).
- ❑ A partnership to get houses for the books built.
- ❑ Stickers to place on the books for numbering.
- ❑ "Little library" boxes or similarly sized boxes so that students can visualize the goal [one for each group].
- ❑ Paper.

Possible Unit Introduction

Students take a field trip to a neighborhood library house that was planted near the school. The students select a book and return back to class to read the book. Students share with classmates about their book. Students return the books. The teacher leads them into a discussion about what they thought about getting a book from the neighborhood library. Inform them that they are going to be able to help create more neighborhood libraries for their community.

Anticipated Questions

- ❑ How do we get the books to the neighborhood?
- ❑ What is a neighborhood library?

- ❏ How many books should we put in the neighborhood library?
- ❏ What is a good book?
- ❏ How many books can fit in a house?
- ❏ When do we take them to the neighborhood?
- ❏ Who would be reading the books?
- ❏ How do I tell someone about a book?
- ❏ Can you write numbers?
- ❏ How do we get library houses to put in the neighborhoods?

Possible Misconceptions/Expected Answers

Lots of students practice counting their numbers before they enter kindergarten. If students attended preschool, they often have learned to count beyond the number 10. However, they may not have truly grasped the pattern of the numbers 0 to 9 and therefore say the wrong number when you start to work on counting from a number that is not one. They are also still developing the connection between the different ways you can "see" the number. For instance, they can say the number two. They maybe can even write the number 2 but they may not be able to hold two objects to represent the number 2. The good news about this grade level is that they don't have a lot of misconceptions. They just don't have a solid understanding developed yet and they just need help with making all of the connections.

Students should be able to count out 20 books for each house that holds the books. Students should be able to properly label each book from 1 to 20. Students should be able to count all of the donated books which should be between 60 and 100 books. They can count by ones or tens. They will write with the support of an adult and technology three or four sentences recommending two to three books.

Formative Assessment

- ❏ **Exit Ticket:** these would be questions related to the current learning experience within the school day. There is a ticket each day. One is for the language arts portion of the project while another is for mathematics.
- ❏ **Quiz:** this is a traditional multiple-choice question quiz that covers at least two standards at a time. This applies for the mathematics and the language arts in the project.
- ❏ **Student Conference:** it is essential for a teacher to spend dedicated time with each student to assess what they know. If meeting with students on a regular basis is not a part of your routine, start with making sure you spend five minutes with students every day.

Reflection Prompts

❑ Did you like your book? If so, why?
❑ How well did you work with your classmates?
❑ Tell me what you know about numbers.
❑ What makes a good book?

Summative Assessment

❑ **Group Neighborhood Library:** this is a group product where three to four students work to prepare one library full of books. Students label books 1 to 20 and place a review of the book in the inside cover. If students are not able to work in small groups, you can have it be an entire class.
❑ **Paired Book Review:** students work in pairs to write a review of the book on a label that will be put in the book.
❑ **Traditional Assessment:** this is a test that contains multiple-choice and written-response questions. It can be created by the district or a group of teachers.

Example Lesson

Mini-Lesson (Whole Class)

1. Make sure everyone is on the mat and ready to participate in the math lesson.
2. Point to the list of questions on the need to know list and where the question *how many books can fit in a house?* is highlighted. Read the question and point to each word as you read. Share how they are going to work on that question during math time today.
3. Move to the section on the project wall that has the math learning objective. Read the objective, *I can count to 100 by ones and by tens*, to the class making sure you point to each word. Then have the students say the objective with you. Thank them for saying the objective with you.
4. Remind them you have been working on counting for a few days now. Have the class practice counting to 20 with you. Count up to 20 a few times as a class.

5. Share with the class they are going to work in small groups with a neighborhood house and see how many books they can fit in the house. They are going to work together.

6. Have a few students join you near one of the houses that has a lot of books near it. Show the group how they will each add a book to the house and count how many as they put it in the house. Stop after you have put three to four books in the house.

Work Time (In Small Groups)

1. Group students off into three or four. Each group should have a house along with 20 to 30 books. Each student in the group should put a book in the house and say the number out loud. They should stop when it is too hard to get books in and out of the box. They write the number of books that fit on a sheet of paper.

2. Students should repeat step one a few times with books of different sizes going into the box. Each time, a different student should write the total number on a sheet of paper.

3. Teacher moves around from group to group asking questions like:
   - How many books do you think will fit?
   - Does the size of the book change how many can go in the box?
   - OH, Scott, that is not the right number after 13. Can anyone help Scott say what number comes after 13?

Debrief (Whole Class)

1. Bring the students back to the mat to discuss the answer to the question. Start by asking students what they noticed about putting books into the house. Use equity sticks (a jar with every student's name on a stick) to draw names to have different students share.

2. Ask students to write on a sticky note how many books they think will fit in a house.

PBL UNIT 2: GEOMETRY (FIRST GRADE)

Subject/Grade Level

Geometry & Informational Reading, First Grade

Common Core State Standards and Mathematical Practices

- ❏ 1.G.A.1 Distinguish between defining attributes (e.g., triangles are closed and three-sided) versus non-defining attributes (e.g., color, orientation, overall size); build and draw shapes to possess defining attributes.
- ❏ 1.G.A.2 Compose two-dimensional shapes (rectangles, squares, trapezoids, triangles, half-circles, and quarter-circles) or three-dimensional shapes (cubes, right rectangular prisms, right circular cones and right circular cylinders) to create a composite shape, and compose new shapes from the composite shape.
- ❏ MP1 Make sense of problems and persevere in solving them.
- ❏ MP4 Model with mathematics.
- ❏ MP5 Use appropriate tools strategically.
- ❏ RI.1.4 Ask and answer questions to help determine or clarify the meaning of words and phrases in a text.
- ❏ RI.1.5 Know and use various text features (e.g., headings, tables of contents, glossaries, electronic menus, icons) to locate key facts or information in a text.
- ❏ RI.1.6 Distinguish between information provided by pictures or other illustrations and information provided by the words in a text.

Time Frame

10 to 12 classroom hours

PBL Question

What is a better design?

Situation/Problem

The world is always working on improving what we currently have available. Generations have experienced new and improved toys. There is nothing that children like more than a toy. Students work in groups to make a better design of the toy of their choice. They read various informational texts to help them understand design and its connection to shapes of various objects.

Materials/Resources

- ❏ Various objects.
- ❏ Construction paper.
- ❏ Clay or playdough.

Possible Unit Introduction

Teacher announces how students will do a brain break and take a little time to play. Students are put into groups where they have access to various objects to play with like blocks and Legos. Students play for about ten minutes. Teacher then shares with students how they get to be toy designers. They are going to work to improve or create a toy that uses common shapes. They will then share these toys with other classes in the school so that they have the ability to have fun when they take their brain breaks.

Anticipated Questions

- ❏ What toy can I redesign?
- ❏ Are there toys people like more than others?
- ❏ How do you design a toy?
- ❏ Which classes get our toys?
- ❏ What shapes are used in toys?
- ❏ How do I read a how-to book?
- ❏ What is a 2D shape?
- ❏ What is a 3D shape?
- ❏ How can I improve my design?
- ❏ When do we give the toys to the other classes?

Possible Misconceptions/Expected Answers

Students are familiar with basic shapes such as triangles, circles and rectangles due to exposure from family as well as previous years in school. However, they are still developing their understanding of dimensions and specifically the words two-dimensional versus three-dimensional. This may be a completely new term depending on if teachers were able to expose them deeply to this concept in previous years. They may not see that a rectangle is two triangles put together.

Students should be able to use vocabulary associated with shapes deeply. They should be able to use words like side, composite, two-dimensional and three-dimensional. They should be able to accurately describe attributes. For example, a triangle has three sides that are closed. They should be able to know that is the case regardless of the color or size. They should be able to put two shapes together to make a composite shape.

Formative Assessment

- ❏ **Thumbs Up/Thumbs Sideways/Thumbs Down:** regularly asking students how well they feel like they understand what they are learning is as easy as them holding out their thumb. They can put it up if they are good, sideways if they feel they know some things but not confident about it and thumbs down if they are lost. This informal assessment that is practiced in most classrooms can happen in projects as well.
- ❏ **Digital Check:** there are so many digital tools that can be used to have students show their understanding quickly. You can use a tool that works like a survey but students use it to respond to a question. You can use district digital tools that allow you to see understanding quickly. You can even gamify the informal check with tools like Kahoot (https://kahoot.com/) or Quizizz (https://quizizz.com/).
- ❏ **Student Conference:** it is essential for a teacher to spend dedicated time with each student to assess what they know. If meeting with students on a regular basis is not a part of your routine, start with making sure you spend 5 minutes with students every day.

Reflection Prompts

- ❏ What shapes are in your chosen toy?
- ❏ What did you learn in this lesson? What are you confused about?
- ❏ How can you tell if something is an attribute?
- ❏ How well did you work with your classmates?

Summative Assessment

- ❏ **Group Redesigned Toy:** this is a group product where three to four students work to prepare one library full of books. Students label books 1 to 20 and place a review of the book in the inside cover. If students are not able to work in small groups, you can have it be an entire class.
- ❏ **Paired Research Results:** students work in pairs to review various informational texts. Students write questions they had before reading the text and answers the text gave them to their questions. In pairs, they write key things they got from the pictures and the text.

Example Lesson

Mini-Lesson (Whole Class)

1. Have everyone at their desks with the following shapes for each group: rectangles, squares, trapezoids, triangles, half-circles, and quarter-circles.
2. Point to the list of questions on the need to know list and where the question *what shapes are used in toys?* is highlighted. Read the question and point to each word as you read. Share how they are going to work on that question during math time today.
3. Move to the section on the project wall that has the math learning objective. Read the objective, *I can distinguish between defining attributes versus non-defining attributes*, to the class making sure you point to each word. Then have the students say the objective with you. Thank them for saying the objective with you.
4. Ask them what word or words they knew in the objective. Call on students to share which ones they know. They may know the words *I, can, between* or *versus*.
5. Ask them what word or words they did not know in the objective. Call on students and have them share which ones they don't know. They will probably say the words distinguish, defining and attributes.
6. Go to the word wall and pull off the words they did not know. Have them repeat the word a couple of times and then review the meaning of each word. Have them repeat the meaning. Ask them if they have a different way of saying the meaning. After a few volunteers add their way of saying it to the meaning. Place back on the word wall and encourage them to come and get the words any time they need it. Remind them that they will do a concept map with the word wall in a few days (see this video from EL Education, an organization that works with teachers

to develop their students, on how to do an interactive word wall: https://vimeo.com/174709417).

7. Share with them that they are going to work together to come up with a list of defining attributes and non-defining attributes for the different shapes. Model how to come up with a defining attribute and a non-defining attribute for triangles.

Work Time (Small Groups)

1. Students work in their groups to come up with at least one defining attribute and a non-defining attribute for each shape. They write their results on a handout that has the shapes and names.

2. Teacher moves around from group to group asking questions like:
   - Does your defining attribute work for all of the squares that are here?
   - How are squares like a rectangle? How are they not alike?
   - Check out the parts of the circle. What do they have that the others don't?

Debrief (Whole Class)

1. Bring the students attention back to you using any method. For example, you can say if you can hear my voice clap once, if you can hear my voice clap twice. Have each group share an attribute and non-attribute for a given shape.

2. Ask students to write on an exit ticket what makes a half-circle, a half-circle.

PBL UNIT 3: MEASUREMENT (SECOND GRADE)

Subject/Grade Level

Numbers and Measurement & Speaking and Listening, Second Grade

Common Core State Standards and Mathematical Practices

- ❏ 2.NBT.B.5 Fluently add and subtract within 100 using strategies based on place value, properties of operations, and/or the relationship between addition and subtraction.
- ❏ 2.NBT.B.6 Add up to four two-digit numbers using strategies based on place value and properties of operations.
- ❏ 2.NBT.B.9 Explain why addition and subtraction strategies work, using place value and the properties of operations.
- ❏ 2.MD.C.7 Tell and write time from analog and digital clocks to the nearest five minutes, using a.m. and p.m.
- ❏ 2.MD.C.8 Solve word problems involving dollar bills, quarters, dimes, nickels, and pennies, using $ and ¢ symbols appropriately. Example: if you have two dimes and three pennies, how many cents do you have?
- ❏ 2.MD.D.10 Draw a picture graph and a bar graph (with single-unit scale) to represent a data set with up to four categories. Solve simple put-together, take-apart, and compare problems using information presented in a bar graph.
- ❏ MP1 Make sense of problems and persevere in solving them.
- ❏ MP4 Model with mathematics.
- ❏ MP6 Attend to precision.
- ❏ SL.2.1 Participate in collaborative conversations with diverse partners about *grade 2 topics and texts* with peers and adults in small and larger groups.
- ❏ SL.2.4 Tell a story or recount an experience with appropriate facts and relevant, descriptive details, speaking audibly in coherent sentences.

Time Frame

12 to 15 classroom hours

PBL Question

How can we put on a great student-led conference?

Situation/Problem

Parent-teacher conferences are regularly scheduled events at every school. As schools moved to more student-centered learning, the conferences changed to student focus. This led to student-led conferences. Students work together to put on a student-led conference. They get a spending limit for refreshments and supplies. They submit a proposed way of spending the money which includes a picture graph. They do a breakdown of the activities that are timed to the nearest five minutes.

Materials/Resources

- ❏ Play money.
- ❏ Digital and analog clock.

Possible Unit Introduction

Students enter the class after attending their specials/electives class. The room has been transformed slightly to an inviting party style set up. The teacher welcomes the students back to class and asks them to take their seat. The teacher shares how for the next few moments they will participate in an informational. The teacher will serve each student a snack. A short presentation on some type of topic students love will be presented for about ten minutes. Teacher will then ask students to share how they liked the informational experience. After students share, the teacher will share how their experience could be a way they conduct the Parent-Teacher Conference that is scheduled soon. Students are informed how they will plan a student-led conference.

Anticipated Questions

- ❏ What is a student-led conference?
- ❏ What is a budget?

- How should we arrange the room?
- What should we serve?
- When is the Parent-Teacher Conference?
- How do we make sure things run smoothly?
- How do we best spend the money we have?
- What is a picture graph?
- What if we go over the money we have?
- What do I want to share with my parents about my learning?

Possible Misconceptions/Expected Answers

Although the students have been working for a few years now on adding and subtracting numbers, students may still struggle with place value and therefore add or subtract the wrong values. They will possibly misunderstand the way to read a hand on an analog clock. They may misuse the symbols for dollars versus cents.

Students are given an amount to spend and create a budget that also includes a picture and/or a bar graph. Students make a plan of the learning they want to share with parents. They practice their delivery of their knowledge before presenting to their parents.

Formative Assessment

- **Entrance Ticket:** these would be questions related to the prior day or days. This can occur each day at the beginning of a whole-class lesson. One is for the language arts portion of the project while another is for mathematics.
- **Recordings:** presenting your understanding to someone else is a great exercise in developing your speaking skills. Using various tools like Flipgrid (https://info.flipgrid.com/) or Voicethread (https://voicethread .com/) allows students to practice speaking while also getting feedback from you and other classmates.
- **Collaboration Checks:** at this stage, students should really have a better grasp on working with people and should be given lots of great feedback on how they are working with people. Use a mailing label document so that you can write feedback to students in the moment and hand them the sticker to place in a collaboration folder.

Reflection Prompts

- ❏ How is an analog and digital clock the same? How are they different?
- ❏ What is one thing you should share with your parents?
- ❏ Sarah has $20 to spend on drinks. She puts four drinks in her basket. They cost $5, $3, $7 and $10. Does she have enough money? Why or why not?
- ❏ What did you contribute to your team today?

Summative Assessment

- ❏ **Group Budget:** this is a group product where three to four students work to prepare a budget for an aspect of the parent teacher conference. One budget can be for refreshments while another is decorations.
- ❏ **Individual Learning Presentation:** students work individually to create a presentation of their learning to their parents. Here is an example of a student-led presentation of learning: https://youtu.be/dmIReiqI1ec
- ❏ **Traditional Assessment:** this is a test that contains multiple-choice and written-response questions. It can be created by the district or a group of teachers.

Example Lesson

Mini-Lesson (Whole Class)

1. Make sure everyone is on the mat and ready to participate in the math lesson.
2. Point to the list of questions on the need to know list and where the question *what is a budget?* is highlighted. Read the question and point to each word as you read. Share how they are going to work on that question during math time today.
3. Move to the section on the project wall that has the math learning objective. Read the objective, *I can solve problems involving dollar bills, quarters, dimes, nickels and pennies,* to the class making sure you point to each word. Then have the students say the objective with you. Thank them for saying the objective with you.
4. Remind them you have been working on counting money for a few days. Have the class say the name and amount for money that you hold up in the air. Remember the money is a dollar and all the coins.

5. Put different coins in the air and ask them how much money you have. Use a random name generator tool to call on different students.
6. Explain to students a budget is an outline of how you are spending your money. Today, they are going to work on how much money they have to spend.

Work Time (Small Groups)

1. Students gather in their group on the mat and are given envelopes which have different amounts of money in it.
2. Students work together to figure out how much money they have in each envelope and write the amount on the outside of the envelope.
3. Teacher moves around from group to group asking questions like:
   ■ What did you do to figure out how much you have? Will you show me?
   ■ When they have written their amounts on the outside, ask them what place the numbers are in.

Debrief (Whole Class)

1. Have different groups share one of their budget envelopes. Ask the students to include how they counted the money.

Appendix 1
Experiencing PBL
Handouts

This section includes the handouts for the Experiencing PBL simulation:

- ❏ Character
- ❏ Counting Our Shopping Items: Make Ten
- ❏ Counting Our Shopping Items: Make Ten Exit Tickets

Downloadable versions of these handouts are available on the book's webpage www.routledge.com/9781646322114.

Name: _____ Date: _____

Character

Pay attention to:

-look
-act
-think
-feel

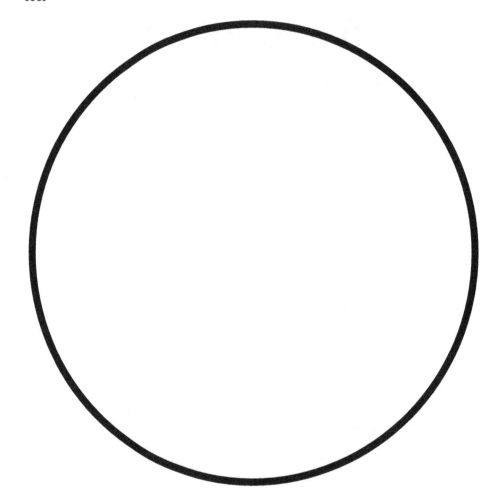

Name: _____ Date: _____

Counting Our Shopping Items: Make Ten
Draw and circle to show how you made ten to help you solve the problem.

1. Kensie brought in 9 pants, and Steve brought in 7. How many pants do they have in all?

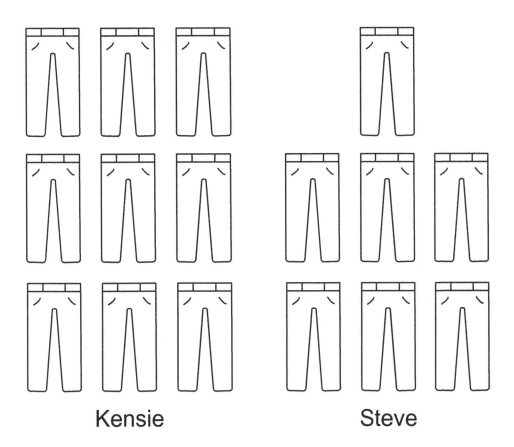

Kensie   Steve

9 and _____ make _____. (9 + _____ = _____)

10 and _____ make _____. (10 + _____ = _____)

Kensie and Steve have _____ pants in all.

2. Lisa donated 9 shirts, and Alice donated 4. How many shirts do they have in total?

9 and _____ make _____. (9 + _____ = _____)

10 and _____ make _____. (10 + _____ = _____)

Kensie and Steve have _____ shirts in total.

3. In the basket, there are 4 scarves. There are 9 scarves in the closet. How many scarves are there all together?

9 + _____ = _____

10 + _____ = _____

There are _____ total scarves.

4. Kensie brought in 9 pants, and Steve brought in 7. How many pants do they have in all?

9 + _____ = _____

10 + _____ = _____

Kensie and Steve have _____ pants in all.

Name: _____ Date: _____

Counting Our Shopping Items: Make Ten Exit Tickets

Draw and circle to show how you make ten to solve.

1. Maria has 3 skirts, and Charles has 9. How many do Maria and Charles have together?

    _____ + _____ = _____

    _____ + _____ = _____   Maria and Charles have _____ skirts.

2. How can this help us count donated items for our shopping day?

Appendix 2
Complete PBL Unit Example Handouts

This section includes the handouts for the complete PBL unit example:

❏ Pirate Project Volume and Weight Record

A downloadable version of this handout is available on the book's webpage at www.routledge.com/9781646322114.

Name: _____ Date: _____

Pirate Project Volume and Weight Record

1. How much treasure?
   a. Weight
      i. Record the values of the following number of pennies:
         1. 12 pennies = __1__ ounce
         2. 24 pennies = __2___ ounces
         3. 60 pennies = _____ ounces
         4. 120 pennies = _____ ounces
         5. _____ pennies = 3 ounces
         6. _____ pennies = 4 ounces
      ii. Record values for 3 other numbers of pennies:
         1. _____ pennies = _____ ounces
         2. _____ pennies = _____ ounces
         3. _____ pennies = _____ ounces

2. How big is my boat?
   a. Volume
      i. Record the length, width and height of your boat using inch cubes or the inches scale of a ruler:

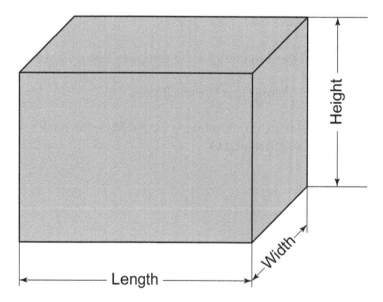

         1. Length = _____ inches
         2. Width = _____ inches
         3. Height = _____ inches
      ii. Multiply the length, width and height to get your boat's volume:
         1. length x width x height = _____

3. Will my boat float?
   a. Can you guess how many pennies will fit into your boat and still float?
      i. GUESS: _____ pennies. Now let's see if you are right.
      ii. Write the number of pennies you estimated and the weights:
         _____/_____*pennies/ounces*
         Example: Pennies/Weight = 72 pennies/6 ounces
      iii. Write your volume: _____ cubic inches
         Example: Volume = 12 cubic inches
      iv. My boat has a volume of _____ cubic inches and can carry treasure as long as the treasure's weight is less than ____ ounces.
      Hint: the volume number for this should be twice the weight number.
      v. Therefore, my boat should carry _____pennies before it sinks.
   Example: therefore, my boat should carry 72 pennies before it sinks.
4. Was your guess correct? _____ [Yes / No]
   a. If your guess was less than the actual number, how many more pennies will you have to add to match your guess? _____
   b. If your guess was greater than the actual number, how many pennies will you have to subtract to match your guess? _____

About the Authors

Telannia Norfar is a mathematics teacher at a public high school in Oklahoma City, OK. She has taught all high school courses, including AP Calculus AB, for over 15 years. As a former journalist and account manager, Telannia found Project-Based Learning (PBL) a viable method for teaching worthy mathematical concepts. She implemented and grew in implementation of PBL in her classroom during her first year of teaching. Soon into her career, she became involved and known by several organizations. She is a board member of the Oklahoma Council of Teachers of Mathematics and member of several organizations, including the National Council of Teachers of Mathematics, International Society for Technology in Education, and Association for Supervision and Curriculum Development. She is a former Teacher of the Year and Rookie of the Year and is a winner of the 2020 Oklahoma Presidential Award for Excellence in Mathematics and Science Teaching. She is a National Faculty member for PBLWorks, where she facilitates PBL training across the United States.

Chris Fancher is a retired Middle Years Programme (MYP) design teacher at a public International Baccalaureate (IB) charter school in Round Rock, TX. He has taught every course, from prealgebra to calculus, in more than 20 years of public education. As a military spouse, he found himself teaching mathematics in four states and two countries, as well as teaching college algebra and college statistics for the University of Maryland University College. Chris honed his Project-Based Learning (PBL) skills at the nationally recognized Manor New Technology High School in Manor, TX. From that experience, he was selected to be on PBLWorks' National Faculty (www.pblworks.org). Prior to the charter school, Chris was an instructional coach for a middle school, where his focus was on helping teachers understand the PBL process and bring their project ideas to life. He was with PBLWorks for six years and facilitated PBL training in 15 states, as well as in China and Australia.

Printed and bound by CPI Group (UK) Ltd, Croydon, CR0 4YY

23/10/2024

01777685-0019